GEOLOGY ON THE MOON

The Wykeham Science Series

General Editors:

PROFESSOR SIR NEVILL MOTT, F.R.S.
Emeritus Cavendish Professor of Physics
University of Cambridge

G. R. NOAKES
Formerly Senior Physics Master
Uppingham School

The Authors

J. E. GUEST is a lecturer at University College, London,
and in addition to his research into lunar and terrestrial
geology, has worked extensively on the geology of Mercury
and Mars as a member of the NASA Mariner 10 and
Viking Orbiter TV Teams.

R. GREELEY works at the University of Santa Clara,
California, and at the NASA Ames Research Center. He
has been closely associated with the NASA Apollo and
Planetology Programs and is a member of the Viking
Orbiter TV Team.

The Schoolmaster

EDWARD HAY is Professor of Geology at De Anza College,
Cupertino, California.

GEOLOGY ON THE MOON

J. E. GUEST
University College, London

R. GREELEY
NASA-Ames Research Center, California

WYKEHAM PUBLICATIONS (LONDON) LTD
(A member of the Taylor & Francis Group)
CRANE, RUSSAK & COMPANY, INC., NEW YORK
1977

Sole Distributors for the U.S.A. and Canada
CRANE, RUSSAK AND COMPANY, INC., NEW YORK

First published 1977 by Wykeham Publications (London) Ltd.

ISBN 0 8448 1170 X

Library of Congress Catalog Card Number 77-15306

Printed in Great Britain by Taylor & Francis (Printers) Ltd.,
Rankine Road, Basingstoke, Hants RG24 0PR.

Cover illustration
Apollo 17 astronaut Harrison Schmitt collecting samples on the Moon.

Preface

Geology began with the recognition that rocks and their included fossils were relics of the past, and that by study of them the history of the Earth could be unravelled. This proved to be no easy task and engendered many long and bitter arguments. For example 18th Century geologists were divided into two camps, the Vulcanists and Neptunists, over the origin of rocks now known to be igneous and not formed by precipitation from sea-water, as some thought. This task of identifying and understanding rock-forming processes still continues to occupy the minds of geologists.

From many individual studies, mainly of small areas, the broad principles of geological science were developed during the latter part of the 18th and early 19th centuries. Progress was slow, hindered by the many hurdles of preconceived notions, but eventually the whole panorama of Earth's history unfolded—history, that is, of the land areas of Earth. It was not until after World War II that man began to explore the 72 per cent of the Earth's surface under the Oceans. With this greater knowledge that was representative of the whole Earth's surface rather than just part of the land masses, a patchwork of local studies were stitched together and combined with geophysical data to formulate a new global approach to earth science known as plate tectonics.

It is perhaps not coincidental that this global view of geology, treating the Earth as a planet, coincided with man's sudden (and to some unexpected) ability to travel in space and look back at the Earth as a globe suspended in space; but this new and exciting concept of place tectonics, a major revolution in geological thinking, to some extent eclipsed that other major geological advance, the geological exploration of the Moon and planets. During the last ten years manned and unmanned spacecraft have photographed the Moon and collected about 400 kg of rock and almost the whole of Mars and about 50 per cent of Mercury have been photographed. Soft-landing craft have successfully photographed the rocky surface of Venus under the shroud of clouds that hides it from outside view, and two Viking craft have landed on Mars. Within a few years the frontiers of geology have been pushed forward to include four other planetary bodies beside the Earth.

Although these geological achievements went unnoticed by many geologists, the return of lunar samples did attract the attention of many highly competent specialists in mineralogy and petrology, and as a result the lunar samples became some of the most intensely studied and analysed rocks ever, which have in six years produced some 30 000 or more pages of scientific literature.

Rocks are the prime concern of geologists but to understand a rock fully its geological environment must be known, and this knowledge on the Moon is provided by photogeology in the absence of extensive field investigations which are thus far not feasible in the hostile lunar environment. The combination of rock samples and photogeological interpretations has made possible great advances in our understanding of the Moon.

In this book we are concerned with what has been learned within the last 15 years about the processes that have produced the present face of the Moon. The principal processes are meteoritic impact and volcanism, two truly planetary processes that have occurred to a greater or lesser extent on all terrestrial planets.

Much of the evidence presented here comes from photogeological interpretation although sample studies are included where they relate directly to understanding surface processes. The extensive and more detailed interpretation of samples relating to their origin, the Moon's internal constitution and the origin of the Moon are only dealt with briefly in this book. The reader intending to take lunar studies further is well advised to read Dr. S. Ross Taylor's excellent book *Lunar Science: A Post-Apollo View* (Pergamon, 1975) which reviews specifically the results of all the work done so far on samples.

The photogeological approach emphasized here is not only an important element of integrated lunar studies including petrology, petrochemistry and geophysics, but also a subject that students of our science can readily study for themselves from the pictures. This book is aimed to provide the basic knowledge to do this.

On the Moon, we have thoroughly tested photogeological techniques by astronaut field studies, sample studies and geophysical experiments. For the other terrestrial planets it is likely that the main bulk of geological evidence will, for many years, come from pictures, and studies of the Moon will stand as a major stepping-stone to geological exploration of the solar system.

Contents

Acknowledgments

This book was the outcome of courses given by us to students at University College, London and Foothill College, California; we hope that these courses, and our book, pass on the excitement and interest of the subject that we have enjoyed as a result of our own research, funded respectively from the U.K. Natural Environment Research Council (JEG) and the National Aeronautics and Space Administration, Lunar Programs Office and Planetary Geology Office (RG).

We are both grateful to our many colleagues involved in lunar and planetary research. Discussions with them and reading their papers has played an important part in developing our ideas. We would particularly like to thank T. E. Bunch, P. Butterworth, D. E. Gault, R. Hawke, J. W. Head III, K. Howard, M. Malin, R. Mason, J. B. Murray, V. R. Oberbeck, P. Schultz, D. Roddy and D. E. Wilhelms, for help in preparing this book. We do, however, take full responsibility for the statements and views expressed throughout. The large—although far from comprehensive—number of references given in the text is an attempt not only to provide the advanced student with further reading, but act as an acknowledgment to scientists who have been responsible for the development of our science in the last few years.

We also thank the ladies who typed the manuscript at various stages in preparation, Jean Cole, Kathleen Beacham and Cynthia Greeley.

Note on Photographs

Wherever possible we have oriented lunar pictures with direction of lighting from the top, top left or left hand side; this should assist the reader in preventing the phenomenon of inverted images in which depressions such as craters appear as hills. We have not been able to do this with oblique pictures, which clearly only have one way up! North is thus not always at the top of the picture.

The photographs of the Moon included in this book represent but a small fraction of those now available from various missions including Ranger, Surveyor, Orbiter, Apollo and Soviet Missions. Catalogues of pictures are available from:

National Space Science Data Center,
Code 601.4
Goddard Space Flight Center,
Greenbelt, Maryland 20771, U.S.A.

for enquirers from within the United States; or from

World Data Center A,
Rockets and Satellites,
Code 601,
Goddard Space Flight Center,
Greenbelt, Maryland 20771, U.S.A.

for those who live outside the United States.

There are now also many maps of the Moon. For a map of the whole Moon, the one published by the National Geographic Magazine is good. There are many others showing smaller regions and for specialized work the new contoured maps at a scale of 1:250 000 prepared from Apollo metric pictures are particularly useful. These are published by the Defense Mapping Agency, St. Louis, Missouri, U.S.A.

1. Introduction

1.1. *A double planet*

The Earth's only large satellite, the Moon, is unique in the solar system because of its large size with respect to its primary planet, the Earth. It is not the largest satellite to a planet, for two of the satellites of Jupiter (Callisto and Ganymede) and one of the satellites of Saturn (Titan) and of Neptune (Triton) are larger; but while these are only a fraction of the size of their gigantic gaseous primary planets, the Moon is almost a quarter the size of the Earth.

The large size of the Moon with respect to the Earth has led to the term *double planet* for the Earth/Moon system. This term is in some ways appropriate as it emphasizes the importance of the Moon as a planetary body; but it should be used with caution as it implies a genetic relationship that does not necessarily exist between these two bodies.

The Moon circles the Earth taking $29\frac{1}{2}$ days in a slightly elliptical orbit bringing it to 364 400 km from Earth at perigee and 406 730 km at apogee (fig. 1.1). With respect to the stars and the Sun, the Moon rotates on its axis; thus there is a day and night for a given point on its surface. But with respect to the Earth it does not rotate and, apart from small variations introduced by such factors as the ellipticity of the orbit (fig. 1.1), it always presents the same face to Earth. We therefore talk of the *nearside* and *farside* of the Moon. Because of these relations the length of one day on the Moon is $29\frac{1}{2}$ Earth-days.

The coordinate system used in this book is known as the Astronautic Convention which on a map puts north at the top, east to the right and west to the left. The earlier convention used by telescopic observers put south to the top but also had east to the right. Both conventions are still to be found, but the Astronautic Convention is the one now used.

1.2. *The Moon's surface conditions*

The Earth has a complex atmosphere giving varying climatic zones each with its own peculiarities, ranging from desert conditions to

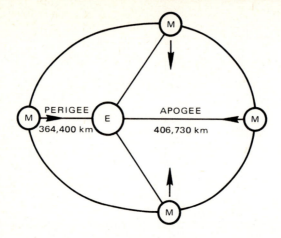

Fig. 1.1. A representation of the Moon's elliptical orbit around the Earth. The Moon (M) rotates so that the same side always faces the Earth (E). The centre of the nearside disc is marked with an arrow showing that in parts of the orbit the nearside does not point directly at Earth (deviation $6\frac{1}{4}°$) allowing an observer on Earth to see parts of the limb region not seen at apogee and perigee. There is also a similar effect in longitude owing to differences in the orbital planes of Earth and Moon; this is known as optical libration and allows us to see, at different times from Earth, 59 per cent of the Moon's total area.

tropical forests. Running water plays an important role in sculpturing the Earth's surface: as well as this, the Earth has a crust which itself is always changing to build mountains, submerge land below the sea, and to fracture certain parts of the surface. The net result of all these factors is a constantly changing face of the Earth.

In contrast the Moon shows little surface activity going on now. There is no water in the free state, and its atmosphere is so slight that all of it could be stored on Earth in a large house at sea level (Bowell, 1971). Essentially the Moon's surface environment may be considered to be in a vacuum, the pressure being only 10^{-4} atmospheres. This slight atmosphere probably consists of the heavier gases, krypton and xenon, although carbon dioxide may have been present for about half the Moon's life. Helium and argon are probably supplied by the solar wind.

Under these conditions, the important processes that erode the Earth's surface cannot exist on the Moon and for this reason alone, its surface is predictably different from that of the Earth.

Temperatures of the rocks at the surface vary according to the time of day: at night they may be as low as nearly -200 °C, whereas at midday they rise to about 125 °C giving a total range of temperature to which rocks are subjected of over 300 °C. However, as can be seen from fig. 1.2, at only 1 metre below the surface the temperature stays constant at 50 °C below zero. The Sun is never higher than $1\frac{1}{2}°$ at the poles and interiors of some craters always remain in shadow. It is

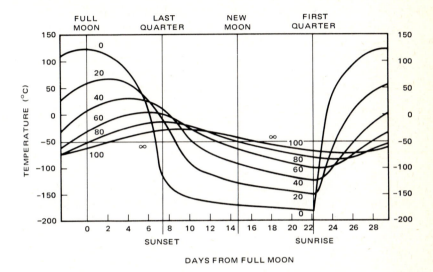

Fig. 1.2. Variations in surface and near-surface temperatures at different times during the lunar day and night. Individual curves represent depths below the surface marked in centimetres. (From Bowell, 1971).

estimated that up to 2 per cent of the Moon's surface is in permanent shadow. As well as possible ice, some workers have suggested that vestiges of the Moon's early atmosphere may be trapped in these regions.

1.3. *Meteorites and the solar wind*

The solar wind consists mainly of protons (hydrogen nuclei) ejected from the sun at average speeds of 500 km s^{-1}. The Earth's surface is protected from these damaging particles by its magnetic field, but the Moon's magnetic field is not strong enough and provides no protection. The damage inflicted by individual impacts of these small particles into the lunar surface is small, but cumulatively they have a noticeable effect on the surfaces of rocks. The protons only penetrate a few millimetres, damaging the crystal structures of minerals in their path. This surface process is akin to weathering in the terrestrial sense.

The main eroding force on the Moon is the steady bombardment by rock fragments of different sizes into the surface at high velocity. On

Earth we see meteors (or 'shooting stars') produced by dust-sized particles burning up in the atmosphere; we also occasionally see fire-balls which are rock fragments large enough not to burn up, but eventually hit the surface. Also the inevitability of the Earth or Moon being hit at times through the geological record by very large bodies, of several kilometres diameter, is strongly illustrated when we re-member that in only the last few decades there have been two near misses. Hermes in 1937 passed within twice the distance of the Moon from the Earth, and Icarus in 1968 passed at only four million miles. Either of these could, on hitting a major city not only have wiped it out, but also devastated a large surrounding area.

On Earth only the larger fragments from space penetrate the atmo-sphere to hit the ground at high velocity. As we shall see later, the velocity is extremely important in determining how much damage is done during impact. A meteorite of mass less than one ton approaching the Earth at an initial velocity of 15–20 km s^{-1} will be slowed down until, at about 20 km altitude, it is falling at its terminal velocity of about 0·1 km s^{-1} (Hawkins, 1964). Only meteorites of mass greater than 1000 tons will impact at cosmic velocities. These larger fragments are rare and do not present a serious hazard. The micrometeorites and meteoric fragments, on the other hand, are numerous and it is estimated that over 1000 tons of this material enter our atmosphere every 24 hours. On the Moon all these particles, large and small, hit the surface at high velocity, peppering it with craters down to the micro-metre size. An area the size of a dinner plate may be hit once per day by a particle 1 μm or larger in diameter. Larger objects become pro-gressively rarer in the Earth–Moon environment and only one impact of an object of mass about 10 g is expected per day in an area of 70 000 km^2. In order to see a crater 1 km diameter being formed, we may have to wait several million years.

1.4.   *Gross features and surface history*

The nearside of the Moon (figs. 1·3 and 1·4) is familiar to most people if only because of the imagined 'Man-in-the-Moon' seen with the naked eye. With the use of a pair of binoculars it becomes clear that the 'face' is made up from two types of terrain. These are the dark, level plains known as *maria* (singular *mare*), and the brighter, densely cratered highlands or *terrae* (singular *terra*).

Marial plains occupy the floors of basins. In some cases these basins are irregular depressions in the lunar surface; but many are deep, circular impact structures surrounded by concentric rings of high mountains. Marial material is generally restricted to the nearside of the Moon, and although the circular basins also occur on the farside they are not filled with marial material to any great extent. This gives the farside a completely different appearance.

Several Apollo missions have collected and brought back rocks representing marial material. It would appear from the petrological studies and isotopic dating that many of the marial surfaces were formed by the extrusion of basaltic lava at between about 3000 and 3700 million years ago. Most of the major surface features including the larger circular basins are older than the maria.

The large circular impact basins are the fundamental structures of the whole lunar surface. Assuming the age of the Moon to be about 4600 million years and the isotopic dates of marial material to be correct, then the circular basins must have been formed during the first 800 million years of the Moon's history (fig. 1.5). Most of the large craters of the highlands also pre-date the maria and must have been formed at this early stage.

After a long period of volcanic activity when most of the maria were emplaced, relatively little happened to change the face of the Moon. There was the steady bombardment by smaller meteoritic bodies, together with occasional major impacts to give the striking rayed craters such as Tycho, Aristarchus and Copernicus. Apart from these the Moon would have looked very much as it does today as far back in time as 3000 million years ago. At the present time, the Moon continues to be bombarded by meteoritic material causing steady but slow degradation. Records of modern impacts have been obtained by seismometers placed on the surface. The seismological records also show that the Moon no longer appears to have much internal activity, although small moonquakes have been recorded.

Thus, compared with Earth, the Moon has been a relatively inactive body for much of its history following the early intense activity of impact and volcanism. This has considerable value in studies of the history of the Earth–Moon system because there are no rocks on Earth that give isotopic dates of older than about 3600 million years. We therefore have no record of the processes operating on Earth for periods before this time. Some scientists have argued that the Moon was responsible for the break in the Earth's record at that time, equating capture of the Moon by the Earth with a hypothetical tidal heating that melted or partially melted the Earth's surface. At the same time, it is argued, heating in the Moon caused melting of sub-surface rocks that were erupted to give the maria. Whatever the cause, coincidence or not, the start of the lunar geological time-scale at about the same time as our record on Earth becomes hazy provides the

Fig. 1.3.   The nearside of the Moon showing selected named features. This is a composite of two half-moon pictures taken with Sun lighting from the east and west respectively. (Lick Observatory photograph).

possibility of extending the geological record back to nearer the time when the Earth, Moon and presumably the other planets of the solar system were formed.

1.5.   *Geological characteristics of the lunar surface*

Craters are the dominant lunar feature on all scales. Several different types of crater are observed, with morphologies that depend on their origin and size. The largest number of craters are less than a few kilo-

Fig. 1.4. Full-moon picture of the nearside. Under these lighting conditions, with sunlight normal to the centre of disc, no shadows are cast to show topographic relief, but the albedos (surface reflectivity) of different units are clearly shown. The highlands are relatively light-toned with numerous large craters and associated bright rays, while the maria are dark and have few superimposed craters. (*Consolidated Lunar Atlas*).

metres across and have a characteristic bowl shape. These are circular in plan, are considered to be of *impact* origin and are peppered randomly over individual surfaces. Young craters of this type have sharp features, older forms have subdued outlines and with increasing age they appear more degraded until they are barely recognizable. The fresher craters have well defined blankets of material ejected from the crater (commonly called ejecta) around them.

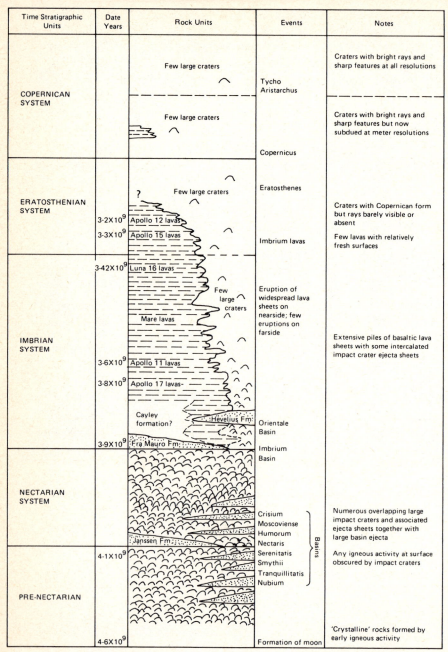

Time Stratigraphic Units	Date Years	Rock Units	Events	Notes
COPERNICAN SYSTEM		Few large craters	Tycho Aristarchus	Craters with bright rays and sharp features at all resolutions
		Few large craters	Copernicus	Craters with bright rays and sharp features but now subdued at meter resolutions
ERATOSTHENIAN SYSTEM	3·2X10⁹ 3·3X10⁹	? Few large craters Apollo 12 lavas Apollo 15 lavas	Eratosthenes Imbrium lavas	Craters with Copernican form but rays barely visible or absent Few lavas with relatively fresh surfaces
IMBRIAN SYSTEM	3·42X10⁹ 3·6X10⁹ 3·8X10⁹ 3·9X10⁹	Luna 16 lavas Few large craters Mare lavas Apollo 11 lavas Apollo 17 lavas Cayley formation? Hevelius Fm Fra Mauro Fm	Eruption of widespread lava sheets on nearside; few eruptions on farside Orientale Basin Imbrium Basin	Extensive piles of basaltic lava sheets with some intercalated impact crater ejecta sheets
NECTARIAN SYSTEM		Janssen Fm	Crisium Moscoviense Humorum Nectaris	Numerous overlapping large impact craters and associated ejecta sheets together with large basin ejecta
PRE-NECTARIAN	4·1X10⁹ 4·6X10⁹		Serenitatis Smythii Tranquillitatis Nubium Formation of moon	Any igneous activity at surface obscured by impact craters 'Crystalline' rocks formed by early igneous activity

Fig. 1.5. A stratigraphic column for the Moon. Rock units shown schematically are: Basin ejecta—dots; crater units—short lunate lines; lavas—horizontal dashes.

Other small craters have different forms: some have no raised rim suggesting that they formed from collapse rather than by excavation; some are asymmetrical or irregular in plan with shallow floors and, from their distribution round large impact craters are interpretated as *secondary impact craters* formed by impact of missiles thrown out from the primary impact crater they surround; a third form of small crater is associated with fault features and is of internal origin, perhaps resulting from volcanic activity.

The larger craters have a different form from the bowl-shaped ones. The most obvious difference is the development of terraces on the inner walls and, in many cases, the presence of central mountains on the crater floors. These craters tend to be relatively shallow compared with smaller craters. Most striking of the large craters are those with extensive bright ray systems which often radiate for hundreds of kilometres. These *rayed craters* are considered to be fresh impact craters formed by collisions of large asteroidal-sized bodies, or comets, with the lunar surface. Studies of craters in general show that with increasing age the rayed craters loose their rays, and with even greater age, continued bombardment by smaller meteoroids degrades their rims until the original crater form is no longer recognizable; all stages in degradation sequence can be illustrated by different craters.

It is clear that a surface dominated by craters will be covered by coalescing sheets of ejecta. The greater the number of craters, the thicker the veneer of accumulated ejecta. This layer of ejected material fragments, that to a greater or lesser extent covers the whole lunar surface, is known as the *regolith*. It consists mainly of local excavated bedrock material although exotic fragments that were thrown from great distances may be mixed in. Most samples collected from the lunar surface come from the regolith. Apollo missions so far have shown this fragmental layer to consist of blocks of bedrock together with *breccias* (coherent rocks consisting of fragments of other rocks) all set in a fine-grained unconsolidated matrix. The fine-grained material consists of finely fragmented bedrock, presumably the end product of repeated impact shattering that has imposed varying degrees of impact shock alteration on the grains. The effects of shock range from simple shattering of the rocks to complete melting which gives glasses that are often present in the form of spherical beads (fig. 8.3). Many breccias have compositions similar to the fines and are essentially indurated ('hardened') accumulations of regolith material that have been welded together by the shock.

Mare ridges or 'wrinkle' ridges (dorsa) are typical features of the maria. They consist of long ridges that can be hundreds of metres wide and tens of metres high. Commonly, they occur as ridge-complexes extending for great distances across the lunar surface and they are thought to be surface 'anticlines' of up-arched surface strata, faults and small lava extrusions from fissures.

The term *rille* (or rima) is an old name for long, trough-like features developed commonly in the maria but also found in certain parts of the highlands. These are fault-bounded troughs known to geologists as *graben*. They are typically several kilometres wide and some hundreds of kilometres long. They may be linear, or, where they occur near the boundaries of circular maria, may be curved in plan view.

In a few places there are prominent normal faults. Some of these may show strike-slip movement but evidence for this is rare on the Moon. Also prominent are fracture patterns that probably reflect close-spaced joint sets.

Another form of rima is the *sinuous rille*. These are in origin different from other rilles, consisting of meandering channels which commonly have a crater at one end. As will be discussed later these features are probably related to the flow of lava.

It is now clear that marial materials consist of extensive lava flows. As well as these extensive lava sheets, there are centres of igneous activity which consist of various constructional forms including domes, some of which are analogous to terrestrial shield volcanoes. Local short, thick volcanic flows also occur. The highland terrain is also made of igneous rocks of a different composition from those of the maria, but the whole highland surface is so modified by impact craters that no obvious surface expression of this early volcanic activity remains.

1.6.   *Origin of the Moon*

Theories of the origin of the Moon may be grouped conveniently into three categories. The first of these, the *capture hypothesis*, has been particularly popular during recent years, and suggests that the Moon was formed in another part of the solar system and was subsequently captured by the Earth. In order for this to happen the two bodies must have had similar orbits. The capture hypothesis presumes that the Moon was formed from the same composition of initial material as the Earth, since they were formed at about the same radius from the Sun. However, if this is so, it is difficult to explain why the Moon is depleted in iron compared to Earth. It should also be noted that, although not condemning this hypothesis, the chances of capture are extremely small.

One of the longest lived hypotheses for the origin of the Moon is the *fission hypothesis*. George Darwin in 1878 suggested that in the early history of the Earth, part of its crust was flung off to produce the Moon, the resulting scar on Earth now being the Pacific Ocean. While on the basis of modern geology, Darwin's idea does not seem reasonable, modifications of the hypothesis have been pursued. It is suggested that as the Earth was separating into a core and mantle, it was rotating in about 4 to 6 hours and thus became so strongly oblate that part of it split away. If this had happened, it is likely that the material thrown out would either have escaped into space completely to give another planet or it would have fallen back on Earth. To get round these objections it has been suggested that Mars was formed in this way (Mars having the correct mass to conform with theory); the Moon would then have been thrown out at the same time and stayed in Earth orbit. Other theories suggest that the Moon was thrown off together with other small offspring and that it was the eventual collision of these smaller objects that produced the larger craters of the lunar surface. O'Keefe (1969) suggests that fission was accompanied by the massive loss of hot atmosphere and that it was from the residue that the Moon was formed. This idea is similar in many respects to a form of the next hypothesis.

The *binary planet hypothesis* involves the Moon being formed as a sister planet to the Earth. This has been the least popular hypothesis because of the difficulty in explaining why the Moon should have developed a different composition from the Earth if they were both formed from the same material. To accomplish the different compositions of the two bodies some fractionation process has to be involved.

This hypothesis has received considerable interest following certain modifications suggested by Ringwood (1970). The modified version is known as the *precipitation hypothesis* in which the Moon is suggested to have formed by accretion of a ring of planetesimals in Earth orbit. The sequence of events postulated is as follows. The Earth condensed from a cold 'cosmic mix' probably having the composition of Type 1 carbonaceous chondrites. This is considered to have been accomplished in less than one million years. As the radius of the Earth grew, gravitational energy of the accreting material increased leading to sustained temperatures at the Earth's surface in excess of $1500\,°C$. This high temperature produced a massive atmosphere, consisting mainly of CO and H_2. Under these conditions it can be shown that silicates would be selectively evaporated into the atmosphere, while metallic iron would continue to accrete on Earth. Also when the Earth's core segregated during a brief period near the end of accretion an increase in temperature would have caused a further evaporation of silicates into the primitive atmosphere.

Fig. 1.6.   An oblique view across the Moon to show typical surface features. A, the 40 km
           diameter fresh rayed crater Aristarchus; B, small bowl-shaped crater; C, crater
           Prinz partly buried by mare lavas; D, mare surface; E, sinuous rille; F, straight
           rille; G, mare ridge; M, pre-Mare materials projecting through more lavas.
           Note how the paths of some of the sinuous rilles have been controlled by fractures
           and graben.

According to Ringwood this atmosphere was dissipated by a number
of factors including intense solar radiation as the Sun passed through
a T-Tauri phase (a short-lived increase in the luminescence), and on
cooling the silicate components were precipitated to form a ring of
planetesimals round the Earth. A further fractionation occurred during

Fig. 1.7.   Apollo 16 view of part of the Moon's east limb and farside. The prominent
dark mare at top left is Crisium, with dark patches of Mare Marginis (near
middle) and Mare Smythii (middle left). The densely cratered nature of the farside
highlands shows well along the terminator where the sun angles are low.

precipitation leaving the less volatile components, these having accreted
into larger objects at higher temperatures and would thus have dissipated less readily. The planetesimals are then thought to have
accreted to give the Moon, which as it grew in size attracted larger
objects eventually capturing bodies large enough to cause the circular
impact basins that form the major structures of the Moon's surface.

Clearly the origin of the Moon is far from solved, but studies of
samples have placed constraints on each of the different hypotheses.

To the geochemist, some form of a double planet hypothesis is accept-able because it is then possible to explain the Moon's chemistry using our present knowledge of the solar system. To the astronomer, capture appears more likely; there are a number of Moon-like bodies in the solar system, but none in orbit around an Earth-like planet. Venus, which in many respects is like Earth and may have been formed in the same way, has no Moon. If the Earth's Moon was captured then Venus's lack of a satellite is no problem, but if some form of binary planet formation occurred it is difficult to see why the Earth–Moon system is unique. Knowledge of the composition of other terrestrial planets will greatly add to our understanding of this fundamental problem of science.

Table 1.1.   Comparative quantities for Earth and Moon

	Equatorial diameter (km)	Surface area (Earth = 1)	Volume (Earth = 1)	Density (kg/m^3)	Surface gravity (Earth = 1)	Escape velocity (km/s)
Earth	12 756	1·000	1·000	5·52 × 10^3	1·000	11·2
Moon	3476	0·075	0·020	3·34 × 10^3	0·165	2·4

2. A geological approach to the moon

2.1. *Introduction*

Anyone familiar with the history of geology will be aware of the bitter controversies that have raged over particular, often fundamental, problems about the Earth. Some have questioned how, if on Earth geologists are not always in agreement (even after painful investigation of problems in the field), satisfactory geology can be accomplished on the Moon. This criticism is valid to some extent in the sense that better geology could be accomplished on the Moon if it were possible to conduct extensive field work, to examine critical rock boundaries, determine lithologies, and collect numerous samples for laboratory work. On the other hand, the Moon, because of its physical properties, does not have a surface that has been formed and modified by such complex processes as operate on Earth, and the task of interpretation is easier. Moreover, there are now available many thousands of photographs of the Moon giving, paradoxically, a better photo-coverage than we have for Earth. This has been made possible by the lack of a lunar atmosphere and accompanying atmospheric conditions that hamper aerial photography on Earth. Lunar photography thus provides ideal material for photogeological investigations.

Sampling and subsequent laboratory analysis of lunar rocks are limited to those returned from six U.S. Apollo sites and the two Soviet Luna (unmanned) sites (fig. 2.1). Most of the landing sites visited were chosen for their geological interest, and the numerous excellent laboratory studies that have been carried out on returned samples have played a critical part in our understanding of the Moon.

Before 1960, the amount of work done on the Moon using truly geological principles was not great. Since then geologists have demonstrated that scientific principles established during a century of geological studies on Earth can be applied to the Moon.

2.2. *Photogeology and geological mapping*

The techniques used to interpret photographs of the Moon taken from orbiting satellites are essentially those well established in terrestrial

Fig. 2.1.   Lunar exploration via successful manned and unmanned landings. Roman
numerals in circles are unmanned U.S. *Surveyor* spacecraft; arabic numbers in
circles are U.S. manned *Apollo* landing site; triangles and squares are Soviet
unmanned *Luna* sites. Spacecraft impact sites are not shown. (From Lunar
Science Institute Map).

geology. There are certain advantages in lunar photogeology. On
Earth, where many different processes of erosion may operate, the
surface morphology depends on the way these processes have acted
on different rocks. Thus the appearance of particular rock types and
the terrain they produce will be different from one area to another. For
example, a granite which weathers into characteristic *tors* in temperate

climates may be altered to a rock known as laterite in the tropics; extensive field work is often required in an area before convincing photogeological maps can be prepared.

This geomorphological problem does not exist on the Moon because external processes operating on the surface tend to be uniform over the whole globe. The main erosive agent on the Moon appears to be bombardment by meteoroids. This bombardment is an almost uniform process over the whole surface of the Moon. Thus it can be argued that on the Moon differences in surface appearance reflect differences in *composition, lithology, mode of formation* and *age* of the materials exposed. This means that different rock units can be distinguished by their surface characteristics with confidence, although their precise lithology, composition and mode of formation must be interpreted. Relative ages can be determined from superposition and geometric relations as on Earth, and extensive units, formed over a short period of time, enable a Moon-wide chronology to be established. The impact crater density on the surface may also indicate the relative age of the unit (fig. 7.11).

There have been several critics (sometimes even geologists) of the value of geological mapping not only for the Moon, but for the Earth as well. Adverse criticism usually results more from genuine ignorance of the methods and aims of mapping rather than from any inherent failings in the technique. The aim of a geologist preparing a geological map is to portray in two dimensions the distribution and three-dimensional geometry of individual rock units of different shapes, sizes and origins at the surface of a planet. Such a map together with a knowledge of the nature of the rocks (and on Earth their included fossils) provides a basis for the understanding of the geological history of the area and the processes by which the rocks were formed. Essentially a geological map consists of a compilation and interpretation of many observations in a form that can be understood by both the author and other geologists; in other words it is a graphical presentation of a large number of data points which presented individually would be difficult to interpret in a disciplined and comprehensible way.

Inevitably, the particular experiences of the geologist will influence the final map: H. H. Read, one of the great proponents of mapping and field work, has pointed out that even national characteristics may have some influence on the final result: for example, the French interpret Alpine structures as a 'series of fluent curves', while the Austrians have 'patterns of straight edged fragments'. To remove as far as possible the subjective element in mapping, rigorous rules for the choice of units that make up the individual elements of the map have been defined. In lunar work the American Stratigraphic Code is followed as closely as is practicable. Once the units to be mapped have been chosen, their boundaries on the map should faithfully reflect their disposition on the ground.

Many geological maps are to some extent interpretative where the mapper wishes to illustrate his own understanding of the way in which a certain structure or group of rocks were formed. To remove such interpretations from a map would be to deny the mapper any scientific inspiration: but if the map is properly constructed any inferences or interpretations will be made clear, and the map should stand as far as the *observational* information is concerned, regardless of the interpretations.

The geological map provides the framework for all other geological and geophysical studies; but perhaps more important it provides the basis for an understanding of the historical development of a planet's surface. Thus on a planet that is heterogeneous, the surface being made up of numerous rock layers of different ages, and possibly different origins, the sequence of these rock units as expressed on the map is our stratigraphical record that provides the history of the planet.

Although simple lunar stratigraphical schemes had been suggested by at least two early geologists, Gilbert and Spurr, no systematic geological mapping of the Moon was attempted until E. Shoemaker of the U.S. Geological Survey set up a branch for lunar and planetary studies. Because it is almost second nature for a field geologist to start an examination of a particular problem by setting down his observations on a map, it is perhaps surprising that no lunar geological mapping had been carried out before this time. The reason seems to be that many studies of the Moon, even those of a geological character, had been done by scientists of disciplines other than geology. It should be emphasized that many of these studies were valuable scientific contributions, but many lacked the cohesiveness that can be provided by mapping.

In a paper that was to be a landmark in the history of lunar studies, Shoemaker and Hackman (1961) proposed not only that the Moon could be mapped geologically, but they also gave a broad stratigraphical time-scale based on their mapping that was to provide the basis for all future mapping, and indeed for our present understanding of the Moon. The place of Shoemaker and Hackman must stand in comparable position to that of William Smith and George Bellas Greenough in the history of geology with their first geological maps of England and Wales. Continued mapping of the Moon by the U.S. Geological Survey has led to expansion and modification of the initial stratigraphical column and the development of techniques to be employed in making geological maps of the Moon and other planets. Many of these ideas are incorporated in an excellent 1:5 000 000 geological map of the Moon's nearside, by Wilhelms and McCauley (1971): this map is strongly recommended to all students of the Moon.

Some emphasis has been given here to geological mapping not only because of its importance as a technique in studying the Moon, but because until recently it has been practised by few workers in the field of lunar geology. It is also probably the best way that a student of the subject can get to know the Moon, and all new students are recommended to try their hand at mapping a small area (figs. 2.2 and 2.3) in order to familiarize themselves with the problems of the lunar surface.

2.3. *Comparative geomorphology*

A high percentage of early work and much modern lunar work has involved drawing analogies between surface features of similar aspects on Earth and Moon (fig. 2.4). The basic concept is that if a feature on the Moon resembles a feature of known origin on Earth, then the two are likely to have had the same origin. This argument has, however, led to many erroneous conclusions in the past because it is often forgotten that more than one process may lead to a feature of a particular appearance (e.g. mudflows and lavas may have similar surface features) or, conversely, that a particular process may lead to different features. It has also been common in lunar studies for features of general similarity to be likened to one another without reference to gross differences in size between them. For example, craters many tens of kilometres in diameter on the Moon have been nonchalantly compared with craters on Earth only a few kilometres across. The worst errors have resulted from making comparisons when the morphological similarities are only superficial; more objective morphometric analysis has later shown them to be quite dissimilar. It is important therefore to employ rigorous scientific arguments when using this technique.

2.4. *Understanding process*

Although the first step towards understanding a lunar feature may be to compare it with a similar feature of known origin on Earth, the investigation should not stop there. Landforms are related to the physical processes that formed them and a proper understanding of the processes is essential to planetary science: with an understanding of the two most important processes that have operated on the Moon, impact and volcanism, great advances can be made by investigations

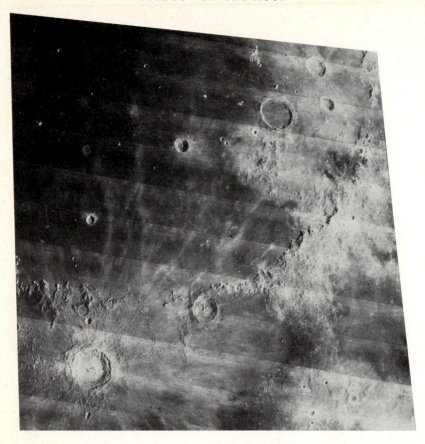

Fig. 2.2.   Southeastern portion of Imbrium in the region of Copernicus and Archimedes. For geological map see fig. 2.3 facing. (NASA Lunar Orbiter IV 126M).

of the geomorphology of landforms produced by these processes. An example is given by cones formed by explosive volcanic activity: will these have the same shape on the Moon as they do on Earth? If the physical principles of explosive ejection are known from studying terrestial volcanoes, then the effect on the shape of the cone of the different surface gravity on the Moon can be determined. Thus on the Moon, where gravity is one sixth of that on Earth, we find that a cone of pyroclastic material (i.e. ejecta from volcanic explosions) formed by a typical strombolian eruption consisting of numerous individual explosions of red-hot lava should have a different shape from its terrestrial counterpart (McGetchin and Head, 1973).

We know more about geological processes on Earth than on any

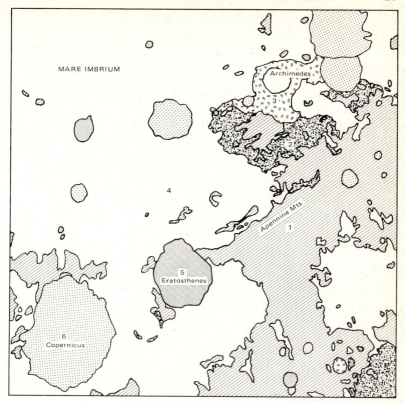

Fig. 2.3.   Simplified geological map of the area shown in fig. 2.2. The units are numbered from oldest (1) to youngest (6). The age relations are determined by superposition and cross-cutting contacts. (After Wilhelms, 1972).

other planet and therefore the more experience the lunar and planetary student has of the Earth the better able he will be to investigate the Moon. However, studies of other planets have tended to show vast gaps in the understanding of our own planet.

2.5.   The quantitative approach

If the physical and chemical processes that operate on planetary surfaces are to be fully understood every attempt must be made to quantify the information available. This is not always easy, but in recent years the whole field of volcanology and impact cratering has been greatly advanced by attempts to obtain quantitative information by field, experimental and theoretical studies.

Apart from the studies of physical and chemical processes there are several fields that lend themselves to quantification on the Moon.

Fig. 2.4.   Similar features on Earth and Moon. The two features on the right are elongate fissure-vents in Hawaii and the Snake River Plains, Idaho. Observations of the Hawaiian structure during eruptions extending over a period of several years show that such features evolve through a series of complex stages of multiple eruptions. Examination of the Idaho structure suggests a similar history. The lunar feature (left) is similar in morphology to these terrestrial structures and this suggests it was formed in the same way.

Investigations of the morphology of lunar features and attempts to compare these with features on Earth are often best done quantitatively. Morphometric analyses of craters has, if nothing else, shown that many lunar craters are quite unlike terrestrial calderas to which they have been likened. There is still much to be done to quantify systematically such things as the sequence of crater-forms produced by degradational processes.

Almost any view of the Moon regardless of scale shows a crater-peppered surface. If we assume that the majority of the craters are of impact origin, the numbers of craters on a particular surface can provide information on the relative ages of different surfaces—the older the surface, the more craters it will have on it. For most geological studies of the Moon it is often a routine matter to assess the numbers of craters per unit area for the individual units, although we will see later that variations in crater density must be interpreted with care, taking into account other geological factors.

2.6. *Other remote sensing methods*

Several non-photographic remote sensing methods can be used to aid geological interpretation of the Moon. Photometry involves measurement of the light-scattering properties of the surface, and can provide information about surface physical properties. Photometric studies show that on the small scale much of the lunar surface consists of loosely packed particles with an effective porosity of 99·5 per cent.

Studies of the way in which light is polarized by the lunar surface can also provide information about its physical properties; the work of Dollfus and Bowell (reviewed by Bowell, 1971) before the manned landings on the Moon showed that the surface had the characteristics of finely fragmented basalt, a result confirmed by collection of actual samples. Both polarization and photometric techniques are less important now that we have lunar samples but the success of these methods for determining lunar surface properties gives confidence to their use for investigations of planets, satellites and asteroids not yet visited by man.

Infrared imaging of the Moon provides interesting information. During the day surfaces absorb heat which is later emitted by radiation. This radiation can be detected in the infrared and is best seen during periods of darkness. The Moon's disc is uniformly heated just before a full-Moon eclipse; at eclipse the heat source is cut off entirely for a short period and differences in rate of cooling can be detected for the whole surface in a single image. Differences in rate of cooling probably reflect differences in the physical properties of the surface rocks.

Various models have been proposed involving different types of roughness of the surface. It is noticeable that 'bright' spots tend to correspond to fresh craters and may result from accumulation of blocks at the surface. Thermal mapping was also carried out from the Apollo 17 service-command module.

At the other end of the electromagnetic spectrum, radar can be used effectively to investigate the lunar surface. There is a correlation between wavelength and the degree of penetration into the surface layer. The longer wavelengths tend to penetrate deeper into the surface and indicate the characteristics of the surface rocks to a depth of several metres. Using improved radar beam resolution, Thompson and Dyce (1966) determined that back-scatter from certain craters was as much as ten times greater than that from adjacent areas. It was concluded that such anomalies appear to correspond with fresh craters. Thus when geological interpretations of individual areas are being made, reference to the radar properties of the surface can be useful.

Radar maps of the Moon bear a remarkable similarity to photographs, but the resolution is poorer. Although these are less valuable when photographs are available, the use of this method to investigate the surfaces of cloud-covered planets such as Venus (fig. 12.10) has already produced pictures suitable for geological interpretation. Infrared and radar maps have shown that lunar craters have associated radar and thermal anomalies that become progressively less noticeable with increasing degrees of crater degradation (Thompson *et al.*, 1974).

2.6. *Apollo landings*

Apollo landing sites were at least in part chosen on the basis of their geological desirability; each site was intended to provide different information from previous ones by landing on units of different age or origin as determined by photogeological interpretation. Ideally a landing site has more than one unit that can be examined and sampled. Each Apollo site was mapped by geologists and traverses were planned to obtain the maximum information from the returned samples and observations.

During the first two missions (Apollo 11 and 12), geological activities were restricted to areas close to the landing craft. The area of study at the Fra Mauro site (Apollo 14) was increased although the astronauts still walked to their sampling sites carrying their equipment in a hand-pulled cart. Apollo 15, the fourth successful landing mission, allowed the astronauts to travel much greater distances using for the first time a battery-driven vehicle called the 'Lunar Rover'; Apollo 16 and 17 used this type of transport and the amount of geological investigation was considerably increased by its use.

The aim of lunar field work is to collect comprehensively from as many different units as possible. Most of the surface is covered by a layer of fragmented rocks derived both from the local bedrock and from elsewhere. In order to understand rocks collected from this layer, it is imperative that each sample is documented before collection, giving its location, orientation on the surface, and the nature of the surrounding ground. Since the bedrock is invariably obscured by the regolith, a means of determining the nature of the bedrock has to be devised to replace direct observation of outcrops. Fortunately many of the craters that pit the lunar surface excavate rocks from levels below the regolith and serve the same function as boreholes would to a geologist on Earth. Study of impact craters has shown that rocks excavated from different levels below the surface occupy different positions in the crater's ejecta blanket. Thus, in order to reconstruct the underlying geology from samples, it is very important to know where, in relation to the crater, samples were collected.

Once Apollo and Luna samples were returned to Earth, they were distributed to more than 150 scientists in many countries. The investigations so far conducted on lunar rocks represent all the possible techniques so far devised to study rocks; each year scientists involved in this work report their results at the annual Lunar Science Conferences in Houston, Texas. Although economic factors have prevented lunar missions subsequent to Apollo 17, the data collected from both manned and unmanned missions continue to yield new information on the geology of the Moon.

Just as important as the samples have been the instruments set up on the lunar surface to study seismic activity, magnetic properties and the solar wind. A network of instruments built up from successive missions has contributed much to understanding of the lunar environment and the nature of the Moon's interior.

3. Circular basins

3.1. *Introduction*

The history of the Moon, especially in its early stages, is punctuated by large-scale catastrophic events. Each of the larger crater-forming impacts threw debris at high velocity for hundreds of kilometres from the crater; and probably few areas on the Moon, during the few minutes after the impact took place, were safe from hypervelocity particles thrown out by the impact. By far the most catastrophic lunar events were the impacts that formed large basins (fig. 3.1). These must have caused Moon-wide disturbances both from falling ejecta (fig. 3.2) and from seismic and shock-wave activity. Rocks surrounding the basin were strongly deformed into large faulted mountain blocks.

Lunar circular basins are essentially large craters and exhibit many of the characteristics found in craters greater than a few tens of kilometres in diameter. It would appear that there is a continuum of structural characteristics from craters to basins dependent on size. Many craters that are less than 150 km in diameter display a mountain or cluster of mountains (sometimes in the form of a ring) rising from the floor of the crater; but between 150 km and 300 km there are always either central mountains or a ring of mountains within the crater. Above 300 km, craters tend to consist of concentric rings of mountains without discrete central mountains—these structures are called circular basins.

Twenty-nine basins, randomly distributed on both the near and farside have been identified on the Moon (fig. 3.3) (Stuart-Alexander and Howard, 1970); those on the nearside are more obvious because they are floored by dark mare materials, whereas on the farside maria are largely absent. There has been some confusion in the past, because some authors have referred to the basins as well as the younger dark lavas as maria; thus the observation that there are few mare materials on the farside has led some to conclude fallaciously that basins are absent on the farside. Here we use the term maria only for the dark smooth plains.

Fig. 3.1. The multi-ringed Orientale Basin. The outer ring is made up of the Cordillera
Mountains (diameter 900 km), while the inner ring is known as the Rook Mountains. For geological map see fig. 3.2. Lacus Veris and Lacus Autumni are small
patches of mare between the mountain rings. The mare region to the top right is
Oceanus Procellarum. (NASA Lunar Orbiter IV, 187M).

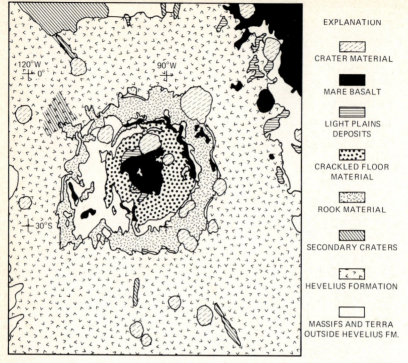

EXPLANATION

CRATER MATERIAL

MARE BASALT

LIGHT PLAINS DEPOSITS

CRACKLED FLOOR MATERIAL

ROOK MATERIAL

SECONDARY CRATERS

HEVELIUS FORMATION

MASSIFS AND TERRA OUTSIDE HEVELIUS FM.

Fig. 3.2. Geological sketch map of the Orientale Basin. (From Howard *et al.*, 1974, after McCauley).

3.2. *Morphology of the basins*

Circular basins, the fundamental structures of the lunar crust, have many features in common with one another and were apparently all formed in the same way. Differences between them include the degree to which they appear to have been eroded (reflecting age differences) and the amount by which they have been buried by later units such as mare lavas. The two youngest major circular basins on the Moon are Imbrium, which is relatively unmodified by later erosion but is covered by substantial areas of mare materials, and Orientale which has not been covered by later units except in a few small areas. All the other basins are modified to a greater or lesser extent but most preserve enough of their original characteristics to illustrate the basic similarities.

Each basin is surrounded by high mountain rings rising up to several kilometres above the surrounding terrain. Depending on the size of the basin there are usually one or more rings of mountains concentric with the main ring. Larger basins such as Imbrium and Orientale, which

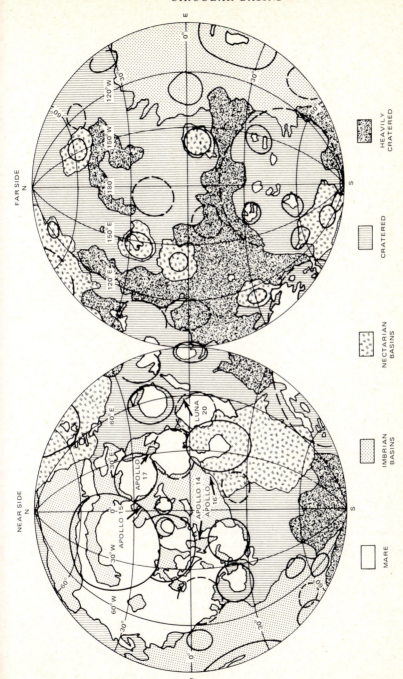

Fig. 3.3. The Imbrium and Nectaris Basin Provinces. (From Howard et al., 1974).

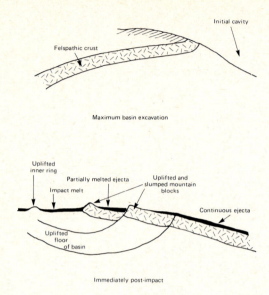

Fig. 3.4.   Diagrammatic profile through a lunar impact basin. The profile on the left
           shows the basin at the time of maximum excavation, with a deep initial cavity,
           and uplift of the surface layers at the crater rim. On the right is the final con-
           figuration of the basin after the floor has uplifted and block-slumping of the rim
           has occurred to form mountain rings and inter-ring terraces covered by ejecta.
           (Modified from Dence and Plant, 1972).

are greater than 900 km across, have two or more rings, whereas
smaller ones typically have only one. These circular mountain chains
tend to be asymmetrical in profile (fig. 3.4) with a steeper scarp facing
into the basin. Between the mountain ranges of fresh basins are level
benches covered by various forms of hummocky material through
which small isolated hills of the underlying rocks protrude. Outside
the basin there are rim deposits that extend for many hundreds of
kilometres and are characterized by intricate ridge patterns. The
mountain rings themselves are often broken into fault blocks.

3.3.   Imbrium and the origin of circular basins

Based on photogeology and Apollo samples, much evidence has been
accumulated that the circular basins were formed by gigantic impacts.
However, in the past the origin of these basins has caused great debate
between those who recognized an impact origin and those who considered the basins to be formed by volcanism and associated tectonic

activity. Because Imbrium is the freshest basin that is clearly visible to telescopic observers on Earth it was the focus of early arguments about the origin of basins.

J. E. Spurr, an American geologist at one time mining engineer to the Sultan of Turkey and later a geologist on the U.S. Geological Survey, wrote *Geology Applied to Selenology* in four volumes (Spurr 1944, 1945, 1948 and 1949) strongly advocating an igneous origin for most of the lunar surface features. He was impressed by the apparent Moon-wide pattern of faults that followed well-defined directions which he called the *lunar grid pattern*, the elements of which trend northwest, northeast and north-south. He noted that patterns associated with the circular basins followed this grid pattern and thus argued that the basins themselves were formed by tectonic forces which caused major subsidence of cylindrical blocks to form the basins. Subsidence was accompanied by crustal uplift to form the circular mountain ranges. Fielder (1965), an English selenologist, has been a modern advocate of Spurr's ideas and has argued 'that there is no evidence whatsoever for a collision in Mare Imbrium' and that 'all the evidence points to the fact that Mare Imbrium is an igneous basin accompanied, in its sinking, by the development of sub-radial and concentric faults'. In support of this view he attempted to show that a number of features such as the radial grooves and sub-radial ridge lineaments were not formed by impact-ejection but are related to tectonism. More important, he argued that sub-radial faults slicing through the ring mountains show movement over a longer period of time, supporting his conclusion that the basin was not formed by a catastrophic impact event but developed by subsidence over a long period of time.

The alternative, now more credible, hypothesis is that the circular basins are gigantic impact craters formed by collision of bodies of asteroidal size with the Moon. The American geologist G. K. Gilbert (1893) found evidence that many of the major crater-like features on the Moon were of impact origin and in his study of Imbrium he drew attention, as Spurr did after him, to the radial pattern of lineaments developed around the basin. Gilbert called this pattern the *Imbrium Sculpture* (fig. 3.5) which he interpreted as ballistic ejection from the impact. Great radial grooves, such as the Alpine Valley (fig. 3.6), were interpreted as furrows cut by large projectiles thrown out by the impact at low angle causing them to skim across the lunar surface ploughing out these features. Baldwin (1963), elaborating on the ideas of Gilbert, recognized several types of radial features including projectile-erosion features and constructional features formed by clots and filaments of molten silicate ejecta.

Fig. 3.5.   An oblique Apollo view of the Imbrium Sculpture northwest of the crater
Ptolemaeus. The Imbrium Sculpture is considered to have formed by scouring of
the surface by Imbrium ejecta. Part of Apollo spacecraft is seen to right. (NASA
Apollo 16–1412).

Modern geological investigations have favoured the ideas of Gilbert
and Baldwin. Geological mapping of the area surrounding basins has
shown that there is an extensive blanket of material distributed round
the whole basin. The surface of this sheet is characterized by some of
the patterns described by Gilbert and others, and the unit is best
explained as a thick rock unit of debris thrown out by the Imbrium
impact event. The presence of this blanket of ejecta forms the mainstay
of the impact argument. The faulting once considered as evidence in
favour of igneous origin for the basin, is now explained as a result of

Fig. 3.6. A Lunar Orbiter view of the Alpine Valley looking southwest into the Imbrium Basin. The mountains adjacent to the mare are known as the Alps and small hills poking through the marial basalts are part of the now buried inner ring of Imbrium. Hilly material in the foreground overlies the mountain material and is thought to consist of a mixture of impact melt and breccia. Note that the Alpine Valley is flooded by mare material and contains a sinuous rille. (NASA Lunar Orbiter V 102M).

the tremendous shockwaves that accompanied the Imbrium event plus post-impact adjustment of the area surrounding this vast hole in the lunar crust.

3.4. *Morphology of the Imbrium basin*

The outer ring of Imbrium mountains (fig. 3.7) is nearly 1300 km across and rises some 7 km above the present floor of the basin. However, the original depth of the basin is unknown because it is now partly filled by mare materials. Baldwin (1963) has estimated that the depth of the mare material is about 5 km, thereby giving a total depth of the basin of about 12 km. This is a relatively shallow depth compared with the original crater of excavation which was probably of the order of 100 km deep at the time of impact excavation; but as we shall see in Chapter 5 it is likely that near the close of impact excavation, the floor of the basin moved forcibly upwards reducing the depth of the crater by many tens of kilometres, and perhaps accompanied the formation of the concentric mountain rings.

Basin mountain rings are characterized by rugged mountain blocks some ten to thirty kilometres across. The blocks have relatively steep slopes although their surfaces are relatively smooth and bright. Wilhelms and McCauley (1971) identify three mountain rings associated with Imbrium, the two inner ones now being largely flooded by mare material. The outer one forms the prominent mountains named the Carpathians to the south, the Apennines to the southwest and the Caucasus to the east. This outer ring is less well developed to the north and west and there is no reason to suggest that this is entirely due to the marial flooding; rather it appears that the mountains were not thrown up as high in these regions as in the south and east.

Of the two inner rings, the one immediately inside the main ring is the best developed and forms the Alps and the mountainous regions near the craters Archimedes and Plato. Little is now visible of the innermost 600 km diameter ring except for smooth hills protruding through the mare plains and mare ridges that form a roughly circular pattern.

Although in some areas ejecta from Imbrium (fig. 3.7) has been flooded by later mare materials, the ejecta forms a well-defined unit surrounding the basin, and has been divided into units according to surface morphology (Wilhelms and McCauley, 1971). Two hilly units occur close to the basin. One of these, referred to as the *Alps Formation*, consists of shock-melted material forming close-spaced, regular-sized hills about 2 to 5 km across. The surfaces are smooth and the hills do not show preferred orientations.

The other hilly unit close to the basin is the *Montes Apenninus material* (fig. 3.8) consisting of both rough coarse blocks that are rectilinear in form parallel to the Apennine Scarp, and smooth to

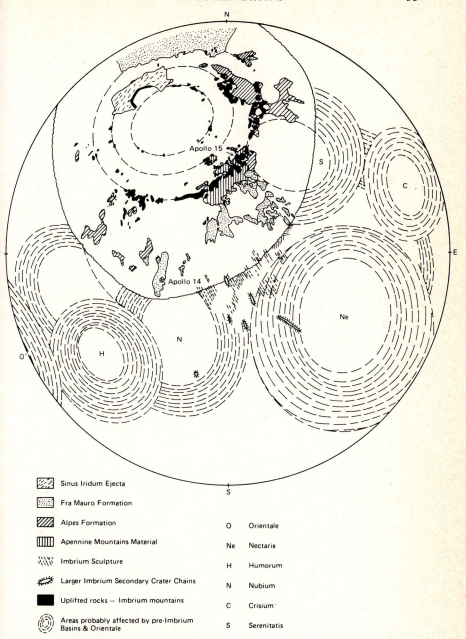

Legend:

- ▨ Sinus Iridum Ejecta
- ▨ Fra Mauro Formation
- ▨ Alpes Formation
- ⫼ Apennine Mountains Material
- Imbrium Sculpture
- Larger Imbrium Secondary Crater Chains
- ■ Uplifted rocks — Imbrium mountains
- ⊚ Areas probably affected by pre-Imbrium Basins & Orientale

O Orientale

Ne Nectaris

H Humorum

N Nubium

C Crisium

S Serenitatis

Fig. 3.7.   Distribution of materials associated with the Imbrium basin on the nearside of the Moon. The approximate extent of materials from other basins is indicated and possible relative ages are shown by overlapping relations.

undulating inter-block materials. This unit crops out on the outer flanks of the Apennine Mountains and also forms benches between the outer and middle ring (well exposed south of Archimedes). The surface morphology suggests that intensely fractured ejecta was dumped on the rim of the basin to form a thick deposit with steep slopes; it was thus unstable and slumped into its present position off the flanks of the mountain during the later stages of basin formation.

The near-basin units grade sharply into the major Imbrium ejecta unit known as the Fra Mauro Formation (fig. 3.9), named after an older crater south of the basin. It was on this unit that the ill-fated Apollo 13 would have landed if it had not run into serious trouble on its way to the Moon; some months later Apollo 14 landed successfully here. The Fra Mauro Formation is characterized by approximately straight, smooth-textured ridges and elongate hummocks 2 to 4 km across and up to 20 km long, each of which is roughly oriented radial to Imbrium. With increasing distance from the basin, the ridged surface becomes braided and the ridges become less prominent as the unit thins. It is clear from the morphology that the ejecta forms a mantle over the underlying cratered terrain, and as the ejecta becomes thinner away from the basin the underlying craters are less subdued and can be seen more clearly as topographic forms. It is the mantling character of·the Fra Mauro Formation that points strongly towards an ejecta origin: if it had been formed by lava flows or ash flows it would have tended to fill in the low ground eventually burying craters rather than being draped over them. In the area around Julius Caesar, a crater some 600 km southeast of Imbrium, Fra Mauro material has been plastered on the walls of the crater facing Imbrium, while on the 'lee' side there is no ejecta in what might be considered a ballistic shadow. Estimates of the thickness of ejecta can be made from the degree of burial of craters and suggest that at the distance of Julius Caesar the ejecta·is about 1 km thick (Morris and Wilhelms, 1967).

At its outer limit the Fra Mauro Formation thins to a feather edge and is difficult to define accurately, and there may be thin deposits outside the normal mappable boundaries. At its outer boundary the Fra Mauro Formation grades sharply into the striking Imbrium sculpture. This terrain consists of well-defined and close-spaced grooves and ridges trending sub-radially from Imbrium, and although Gilbert originally thought that it had been formed by scouring of the surface by ejecta, many recent writers have thought it to be close-spaced faulting. However, the high resolution photography from Apollo shows that the grooves have scalloped edges and are more like long lines of craters. This observation together with the greater understanding of the impact mechanism has swung the general opinion back to the ideas of Gilbert and it now appears probable that the Imbrium Sculpture was produced by a rain of high velocity ejecta particles excavating the surface: an interpretation that is further enhanced by the presence of large numbers of secondary craters associated with, and outside the range of, the sculpture. These are irregularly shaped craters, often elliptical in plan view, with their long axis pointing towards Imbrium to form chains. They are considered to be produced by long-range missiles from the basin hitting the surface at high velocity (but below escape velocity) and excavating craters. As we shall see in the next chapter this is a normal characteristic of impact cratering. Secondary craters can be large and at 1300 km from Imbrium the maximum diameter of secondary craters is about 9 km.

Probably associated with Imbrium's formation was Moon-wide seismic activity shattering the Moon's lithosphere and imposing on the Moon a well-defined pattern of fractures that were either radial or concentric to the Imbrium basin.

In summary, the photogeological evidence from Imbrium can be used to provide an idea of the mechanism of formation and the sequence of events that occurred. It should be remembered that crater excavation probably lasted only a few minutes and the whole event was probably complete within an hour or so. To a geologist accustomed to rock sequences and structures that took many thousands or even millions of years to develop, such a catastrophic event that produced high mountain ranges and sequences of rocks several kilometres thick within the span of only minutes or hours necessitates reconsideration of conventional geological ideas.

3.5. *Evidence from Apollo landings*

Because of its wide areal extent and rapid emplacement as an ejecta sheet, the Fra Mauro Formation is an important geological unit to collect samples from. The first attempt to land on Fra Mauro material failed, owing to a nearly disastrous explosion on the Apollo 13 spacecraft during its flight towards the Moon. Following this, Apollo 14 was re-scheduled to the same site and on 5 February 1971 astronauts Shepard and Mitchell landed on the Fra Mauro Formation near its type locality some 550 km south of the Carpathian mountains. Geological mapping of the site shows that it consists of hummocky terrain with irregular hills and broad undulations together with linear ridges some 100 metres high forming part of the sub-radial pattern characteristic of Fra Mauro.

In order to sample from one of these ridges the astronauts walked

Fig. 3.9. Oblique Apollo view of the Fra Mauro area near the Apollo 14 landing site
(indicated by arrow). Fra Mauro material to the left of the picture is seen overlying older craters which have been modified by Imbrium Sculpture. Part of
Apollo spacecraft seen to right. (NASA Apollo 16–1418).

(*Opposite*)

Fig. 3.8.   The area southeast of Imbrium. Near the top left-hand corner are the Apennine
Mountains; immediately southeast is ejecta of the Apennine Mountain type
which presents a ridged upper surface. The Apennine material merges outwards
into typical Fra Mauro material with longitudinal ridges seen in the lower right
of the picture. All these units have been partly overlain by dark mare material.
Sun lighting is from the right, with north at the top. (NASA Lunar Orbiter IV
109H$_2$).

from the lunar landing module to the top of a nearby ridge where a fresh crater (named Cone Crater) had excavated deep into the ridge. Near the lip of the crater the astronauts found the terrain to be very blocky with boulders up to 10 m across, and here Sheppard and Mitchell got lost! Only later was it discovered that they were within 20 metres of their objective, the crest of the crater rim, and the samples that they collected are considered to represent rocks from deep within the ridge.

Samples brought back from the Fra Mauro Formation consist of impact breccias showing different stages of shock metamorphism. This was good confirmation of the photogeological interpretations of the Imbrium Basin. The breccias contain fragments of a variety of rock types and fragments of minerals, set in matrices that range from friable and fine-grained to coherent mosaics of euhedral (well formed) crystals. Most of the clasts are angular but rounded clasts are not uncommon. Boundaries of the clasts may be sharp or diffuse, merging into the host matrix. Up to three generations of breccia clasts have been noted showing that the breccias have a complex history, breccias being re-brecciated by later impacts. Thus we are probably seeing evidence of several brecciations that may be both pre-Imbrium and post-Imbrium, as well as that produced by the Imbrium event. The friable breccias are probably post-Imbrium and result from local small impacts on the Fra Mauro surface. More coherent breccias either were produced by the Imbrium event or are pre-existing breccias excavated from the Imbrium Basin. In order fully to understand the Apollo 14 rocks, some distinction between Imbrium and pre-Imbrium breccias must be looked for, but at the present time this distinction is not clear. Dence and Plant (1972) have argued that from our understanding of impact cratering most of the material excavated from Imbrium would not have been shocked or raised in temperature to a very high degree. They suggest that much of the annealed, recrystallized breccias from Fra Mauro are breccias from pre-existing major impacts and that the effect of Imbrium shock-metamorphism at this distance (500 km from the margin of the basin) was to produce only weakly shocked rocks. Following this interpretation the multiple shock and thermal events recorded in the fragmental rocks are probably the result of intensive impact bombardment *before* the Imbrium event. However it could be argued that the variations in shock lithification result from layering in the Fra Mauro unit, the lower layers having remained hotter to become more annealed.

A date for the Imbrium event can be obtained from Apollo 14 samples. Rb-Sr and K-Ar ages for the clasts suggest an age of 3900 million years as the age of the rocks incorporated in the Fra Mauro Formation; since the oldest mare lavas so far dated are about 3700 million years, the Imbrium event must lie between these dates and probably close to 3900 million years. All the largest basins except Orientale are older than Imbrium, indicating that they were formed during the first 700 million years of the Moon's history.

The second mission to Imbrium was Apollo 15 when astronauts Scott and Irwin touched down near the foot of the Apennine Mountains in Palus Putredinis just inside the basin (fig. 4.8). This was the first mission to use a Lunar Rover and the astronauts travelled considerable distances from their landing craft. They visited the lower slopes of the Apennine Mountains to collect rocks from the base of the mountains as well as material from higher up that had rolled down the mountains that rise some 3 to 4 km above the site.

Among the rocks collected were breccias of several different types. Many of the breccias were well lithified with abundant granulated feldspar clasts and an aphanitic (fine-grained) matrix which sometimes intrudes the clasts. Coherent breccias with vitreous groundmasses are also present containing basalts together with granulated olivines and pyroxenes. Presumably both these breccia types represent fragmented bedrock; but there are also friable regolith breccias which on the basis of their constituent clasts (consisting of non-mare basalt older than the basin, as well as basalt from the nearby marial surfaces and glass fragments) are considered to represent reworking of the surface at a much more recent date (see Chapter 8).

The coherent and lithified breccias representing the mountain bedrock show that the lunar surface rocks into which Imbrium was excavated consisted of fractionated igneous rocks of a different composition from later marial basalt. Anorthositic rocks (essentially rocks made only of calcic plagioclase) are particularly prominent and suggest that they are an important constituent of the early lunar crust.

The brecciated nature of the bedrock here implies that it was brecciated even before the Imbrium event, an observation to be expected as the bedrocks uplifted in the Imbrium event were ejecta from the earlier Serenitatis basin adjacent to Imbrium (fig. 1.3).

3.6. The Orientale basin

Orientale is the youngest major basin on the Moon (fig. 3.1) and is the freshest known impact basin in the solar system (Moore et al., 1974). Apart from a few small patches of mare and a few large impact craters almost all of its original surface is still exposed. The basin itself has a diameter of about 900 km and lies on the western limb of the Moon (this is the eastern limb following the early *astronomical convention* and hence its name). The basin is only just visible from Earth at suitable librations and even then little of it can be seen because of the extreme foreshortening. Its full extent was first shown by Lunar Orbiter photographs which provide good information for the eastern side. Unfortunately during all the Apollo missions the basin was in darkness and our studies must rely almost entirely on Lunar Orbiter data.

As with Imbrium there are several different types of ejecta associated with the basin (Head, 1974; Moore *et al.*, 1974) depending on distance from the basin centre (fig. 3.2). In the middle of the basin, within the Rook Mountains and partly covered by mare materials, is a rock unit whose surface is corrugated ('crackled floor material', in fig. 3.2). The surface is pitted and cut by cracks ranging in size from below the resolution of the pictures to 2 km across (fig. 3.10). The corrugated surface is thought to be cooled ejecta-melt draped over a pre-existing hilly topography; in some places these hills protrude through the ejecta as 'islands'. The general surface form and texture suggests that it was formed by hot material that cooled as a veneer over an irregular surface, developing cooling cracks and tension features as the hot system settled and cooled. Associated with this unit are smooth plains that grade into the corrugated unit. These plains are also cut by similar cracks and furrows to those on the corrugated areas. In general the cracking is radial and concentric to the basin except where it forms patterns associated with the underlying terrain.

Impact melt in the middle of the basin is to be expected from our knowledge of impact cratering mechanics (see Chapter 5), which suggests that the majority of impact melt will stay as a pool within the crater rather than being excavated. This relation is found in terrestrial impact craters such as Manicouagan and the Clear Water Lakes (Dence, 1968). Based on this and photogeological evidence for Orientale it is expected that the rock making up this unit would have congealed from varying proportions of partly melted material and non-molten rock. The most liquid melt probably ran off the topographic highs to give smooth plains (Head, 1974), and drained from beneath crusted melt into the centre of the basin (Greeley, 1976). Material of the corrugated type is also found on the floors of larger craters such as Tycho, Aristarchus and Copernicus, supporting the view that craters and basins were formed in the same way.

Between the Rook and the Cordillera Mountains is a hilly facies (Rook Material, figs. 3.2 and 3.11) characterized by close-spaced dome-shaped hills about 1–5 km across, generally circular in plan. The hills tend to be random in distribution although in some places

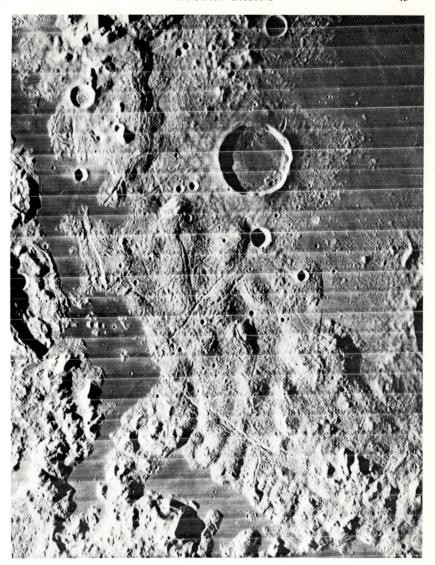

Fig. 3.10. Orbiter view of the inner part of Orientale. In the top right is dark mare material filling the inner part of the basin. Patches of mare also occur around the foot of the Rook Mountains (bottom left). The middle of the basin is crackled floor material interpreted as impact melt and breccias. In the upper middle part of the picture is the 35 km crater Kopff; note the other young crater Maunder of similar size (cut by the right hand margin of the picture) with well developed ejecta facies. Also note graben cutting crackled floor material but overlain by mare material. (NASA Lunar Orbiter IV 187H$_2$).

they are aligned sub-radially to the basin. This material is generally considered to have been emplaced as hot ejecta similar to the corrugated facies but probably with a greater mixture of cold material.

The main ejecta of Orientale (Hevelius Formation, fig. 3.2), lies outside the Cordillera Mountains (fig. 3.11). Close to this mountain range the surface is broken into long ridges and valleys roughly parallel to the mountain crest; but with distance from the mountains this concentric pattern becomes chaotic. The concentric pattern suggests that there was extensive slumping of ejecta off the Cordillera Mountains and away from the basins. Circular depressions may indicate the presence of buried craters.

Concentric facies* grades outward into ejecta characterized by radial ridges and valleys forming a braided pattern similar to that of the Fra Mauro Formation. This pattern continues for as much as 300–400 km from the Cordillera Mountains, but with increasing distance from the basin the surface pattern becomes smaller in scale and less pronounced. In the thinner parts of the ejecta sheet, buried craters can be seen and the patterns on the ejecta are modified around them in the form of banked ejecta on the up-range walls of the crater (fig. 3.12), and dune-like features inside the crater. The front of the ejecta sheet appears to be lobate. All this evidence suggests that during emplacement the ejecta sheet moved forwards for considerable distances as a surface flow of fragmental material, the ejecta wrapping around the walls of craters, and developing flow, slump and avalanche surface characteristics.

Outside the continuous ejecta the pre-Orientale terrain has been cut, grooved and pitted by secondary impact craters occurring as discrete craters, crater chains and clusters. Also associated with the secondary cratered areas are grooved and lineated terrains. These regions were strongly modified by debris thrown out from the basin at high velocity cratering and grooving the surface; there was not sufficient material at this range to form a continual sheet of ejecta.

In summary the following sequence of events is suggested for Orientale:

1. Impact of a large extra-lunar body into heavily cratered terrain, to form a deep transient cavity, of uncertain diameter, but which may correspond to the Rook Mountains.
2. Molten and partly molten shocked materials lined the interior of this cavity.

---

*Individual rock units such as sheets of ejecta often show lateral variations owing to different modes of formation in different areas. These different parts of the unit are known as *facies*.

Fig. 3.11. Orientale ejecta near the Cordillera Mountains. The hummocky material in the lower right is Rook material inside the Cordillera Mountains. To the middle and top left is swirley ejecta of the Hevelius formation. (NASA Lunar Orbiter IV 180H₃).

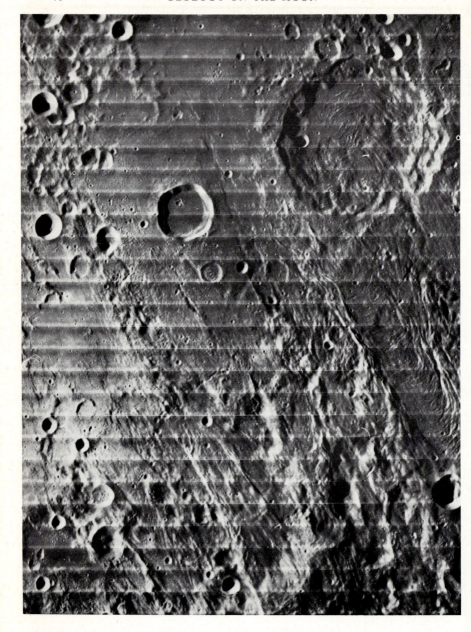

Fig. 3.12.   Orientale ejecta in the outer parts of the continuous ejecta (Hevelius Formation) showing 'flow' markings, and swirls inside a crater at top right. (NASA Lunar Orbiter IV 172H$_2$).

3. Ejection of high velocity material that landed far out from the basin giving secondary cratering.
4. Ejection of increasingly lower velocity ejecta that landed nearer the crater.
5. Continued movement of ejecta away from the crater by radial flow.
6. The floor of the transient cavity rebounded up to its present position, the basin slumped inwards producing the large fault inside the Cordillera Mountains; and the molten ejecta inside the basin settled and cooled to give the present topography. It is estimated that the total volume of melt material produced during the Orientale event was in excess of 200 000 cubic kilometres. This agrees with theoretical calculations for the amount of melt and would imply by extrapolation that the Imbrium Basin produced over a million cubic kilometres of melt (Head, 1974).
7. Production of small volumes of mare basalt to form Mare Orientale, Lacus Veris and Lacus Autumni.

3.7. *Other basins*

Although none of the other major circular basins greater than 400 km across are as fresh as either Imbrium or Orientale, they are generally recognized by remnants of concentric mountain rings and traces of the original ejecta sheet. For example, there is a well-defined continuous ejecta sheet, named the Janssen Formation, south of Nectaris. This unit has well-developed sub-radial ridges and is analogous to the Fra Mauro Formation. It is likely that the pitted and ridged terrain of the Descartes Mountains sampled by the Apollo 16 astronauts is ejecta from Nectaris.

It is clear that the large basins are the fundamental structures of the lunar surface. Stratigraphically all regions of the Moon are underlain by thick sequences of superimposed basin ejecta; these sheets are complex because there has been a great deal of mixing of ejecta from different events. Close to the basins the crustal rocks are strongly deformed by the uplift of the mountain chains along major faults, together with overturning and thrusting; further from basins (particularly from Imbrium) the lunar crust is intensely fractured giving patterns that have been utilized by later tectonic movements.

4. The maria

4.1. Introduction

Since man first turned his questioning eyes to the Moon, there has been speculation about the dark, smooth areas—the maria (fig. 4.1). The term maria in itself reflects early beliefs that they were lunar oceans. Through the years, with improvements in telescopes and remote sensing methods and advances in knowledge of geological processes, the origins for the marial plains gradually narrowed to a few possibilities, including water-laid sediments, impact melts, dust deposits and volcanic rocks. However, even before Apollo it had been generally considered that the maria consisted of sheets of volcanic rock. From the way that the maria embay the highland terrain on their margins, it was argued that the maria consisted of younger rock units filling pre-existing hollows (fig. 4.2). Because of the lack of running water or wind to transport debris from the highlands to the lowlands, volcanism was the only known geological process that could have emplaced the maria.

Telescopic observations also showed a number of features in the maria that resembles volcanoes, and, under low lighting conditions, lobate scarps resembling lava flow fronts were recognized in Mare Imbrium by Robert Strom. Despite the increased resolution of surface features gained from unmanned space-probes, and chemical analyses by landed Surveyor spacecraft all indicating volcanic materials, the matter was still not resolved in the minds of some lunar scientists.

Samples of marial rocks returned from manned Apollo landings and unmanned Luna missions have consistently revealed basalts and there is no doubt that the marial plains are vast lava flows that have flooded low-lying regions of the lunar surface, primarily on the nearside.

Although until samples were examined, the geological ages of the lunar maria were thought to be fairly young, it is now known from radiogenetic dating that the *youngest* mare basalt sampled so far is near to 3000 million years old. While there are, on the basis of crater counts and analysis of surface morphology, younger mare units than those sampled, it is doubtful that even the youngest lunar basalts of any appreciable extent are much younger than 2800 million years.

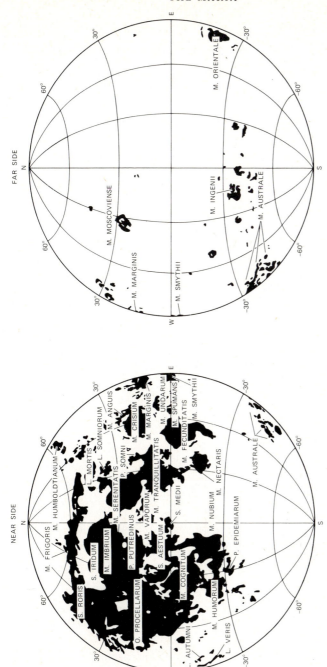

Fig. 4.1. Distribution of mare materials. (After Head, 1976.)

Fig. 4.2.   Hadley Rille in Mare Imbrium. This 200 m wide rille starts at the foot of the
Apennine Mountains and was probably formed as an open lava channel during
the emplacement of the surrounding mare lavas. Some sections may have been
bridged to form lava tubes. North is at bottom. (See fig. 4.8 for map). (NASA
Apollo 15–0587.)

In this chapter, we discuss the general nature of mare surface features
and the characteristics of mare basalts as derived from returned
samples. Our discussions of the lunar samples are not detailed and the
reader is referred to the excellent treatment presented by Taylor
(1975) for a general synthesis of the lunar sample studies.

4.2. Marial basalts

Samples of marial material (fig. 4.3) have been obtained from Mare Tranquillitatis (Apollo 11), Mare Cognitum (Apollo 12), Mare Fecunditatis (Luna 16), Mare Imbrium (Apollo 15), and Mare Serenitatis (Apollo 17). Although all the marial rocks from these sites are essentially basalts there are certain differences in texture and composition which can be related to differences in origin and the processes leading up to their crystallization.

The most abundant minerals in the marial rocks are essentially the same as those of terrestrial basalts and gabbros, namely: pyroxene, plagioclase, ilmenite, olivine, and spinel. A great many other minerals and mineral assemblages have been identified, including several new minerals apparently unique to the Moon such as pyroxferroite, tranquillityite and armalcolite. Notable for their absence are hydrous minerals—thus answering the question of the availability of water on the Moon. Their lack confirms the long-held contention of many investigators that water (in any state) is absent on the Moon and that, throughout its history, the Moon has never had water present in substantial quantities, if indeed in any quantities at all (the exception is the single occurrence in the highlands of goethite, $FeO \cdot OH$).

By far the most abundant minerals in the lunar basalts are the *pyroxenes*, of which the two major varieties are pigeonite (low calcium) and augite (high calcium). The generalized chemical formula for pyroxene is XYZ_2O_6, in which for lunar pyroxenes X = Ca, Y = Mg, Fe, Ti, Al, Mn, and Cr, and Z = Si and Al. Detailed studies of the chemistry and crystal structure of the lunar pyroxenes provide clues to the geochemical history of the rock in which they occur.

Plagioclase feldspar is the second most abundant mineral in the samples, as would be expected in basaltic rocks. Specifically, Ca-rich members of the plagioclase group predominate.

Ilmenite ($FeTiO_3$) is the third most abundant mineral of the lunar rocks, although the amount is variable among the mare sites. This variability suggests some important differences in the physical behaviour of the flows during their emplacement, as discussed below.

The amount of *olivine* $(Mg, Fe)_2SiO_4$ also varies depending upon locality. In general it is the fourth most abundant mineral. In some rocks, olivine occurs as phenocrysts (larger crystals among small ones) of sizes up to 2 mm and in general the olivine began to crystallize early. Most of the olivine has a fresh appearance and the crystals are only slightly altered. Considering the age of these rocks their freshness is startling compared with terrestrial rocks.

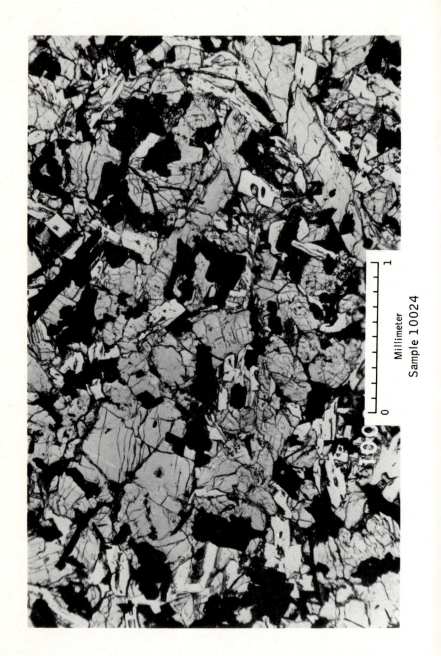

Sample 10047

Millimeters

0   1   2

Fig. 4.3.   Photomicrographs of Apollo 11 lavas. The principal minerals are plagioclase, pyroxene and opaque ilmenite. (NASA pictures.)

The *spinels* constitute a large group of diverse minerals having the general formula AB_2O_4 (where on the Moon $A = Fe^{2+}$, Mg, or Mn; $B = $ Al, Cr, or Ti). Individual minerals include chromite ($FeCr_2O_4$) and spinel ($MgAl_2O_4$).

Pyroxferroite was the first new mineral described from the mare samples and is an important constituent of the microgabbros and gabbros. In fact, lunar pyroxferroite contains about 6 per cent CaO with minor amounts of Mn, Mg, Ti, and Al, resulting in a rather complex formula. It occurs only as a late stage accessory mineral, often in intimate growth-relations with augite. From petrographic studies of associated minerals and studies of laboratory simulations, the mare pyroxferroite must have crystallized from an iron-rich residual liquid which was cooling rapidly.

Armalcolite is another of the unique lunar minerals that was discovered in the samples returned by Apollo 11. Its name is derived from the Apollo 11 crew: *Arm*strong, *Al*drin, and *Coll*ins. It is a rather rare mineral that is opaque, grey in reflected light, and nearly always rimmed by ilmenite. The name is applied to a range of chemically similar minerals lying between two end-members $FeTi_2O_5$ and $MgTi_2O_5$. From studies of its texture and laboratory syntheses, armalcolite appears to be one of the early minerals formed during cooling of the lunar lavas; when the temperature fell to about $1130°C$, it reacted with the remaining melt to form ilmenite. The small amount of armalcolite remaining today represents remnants of the mineral that did not completely react before the lava solidified.

Numerous other minerals occur in marial materials, including tridymite and cristobolite (both SiO_2).

There is a wide range of textures among the lunar basalts. In most cases, there textures reflect the environment of formation; however, often samples from a single site may display a divergence of textures and it is difficult to envisage such a wide range of conditions in such a restricted area. Since the samples are loose rocks collected from the surface, they may be representing completely different flow units which have been excavated from different depths by impact.

Textures of lunar basalts range from vitrophyric to coarse-grained, and include porphyritic rocks. Typical textures are:

vitrophyric — phenocrysts (commonly olivine and/or pyroxene) in a conspicuous glass or devitrified glass groundmass.

variolitic — radial or sheaflike aggregates of crystals (often plagioclase) in a fine-grained groundmass.

subophitic — plagioclase intergrown and partially enclosed by subhedral pyroxene and olivine.

intersertal — network of plagioclase, pyroxene, olivine, and ilmenite phenocrysts set in a groundmass of small crystals and glass.

gabbroic — coarse crystalline (>2 mm).

Like terrestrial basalts, many of the lunar basalts are vesicular (full of holes caused by expansion of gases).

In general, the major element chemistry of the mare basalts is similar to many terrestrial basalts except that lunar basalts are strongly reduced and the Fe:(Fe + Mg) ratio is higher than most basalts on Earth, while the Ti in Apollo 11 basalts is higher than any terrestrial basalts. Table 4.1 shows the typical chemical composition of the range of lunar basalts. The trace and minor element chemistry of the mare basalts is unique, being depleted in volatile and noble elements and enriched in the refractory and rare earth elements, which provides clues to the geochemical origin and history of not only the mare basalts, but the entire Moon as well.

The textures and isotopic ages leave little doubt that all the mare basalts crystallized from internally generated lavas. There is however, much debate over the degree to which their compositions have been been affected by near-surface fractionation processes. Thus, it is not yet clear which, if any, mare basalt compositions represent primary unmodified melts from the lunar interior.

The mare basalts can be grouped broadly into two sets, the high-titanium basalts and the low-titanium basalts. High-titanium basalts include the ilmenite basalts of Mare Tranquillitatis (Apollo 11), and at least the older basalts of Mare Serenitatis (Apollo 17). Low-titanium basalts include the younger basalts from Oceanus Procellarum (Apollo 12), Mare Imbrium (Apollo 15) and Mare Fecunditatis (Luna 16). Although many investigators further subdivide these groups, for purposes of our discussion, we will restrict ourselves to this broader classification.

Commonly, the low-titanium basalts are porphyritic and show chemical variations that can be explained by low-pressure fractionation, which would suggest near-surface crystallization. In addition, some members of the group appear to have existed at the surface as liquids. From various model analyses and phase-equilibrium studies, many workers have concluded that the low-titanium mare basalts are derived by partial melting of olivine pyroxenite at various depths ranging from 150–400 km (Hays and Walker, 1974).

Table 4.1. Typical analyses of marial rocks. (From Taylor, 1975.)

	Green glass Apollo 15	Olivine basalt Apollo 12	Olivine basalt Apollo 15	Quartz basalt Apollo 15	Quartz basalt Apollo 12	High K basalt Apollo 11	Low K basalt Apollo 11	High Ti basalt Apollo 17	Aluminous mare basalts Apollo 12	Aluminous mare basalts Luna 16
SiO_2	45·6	45·0	44·2	48·8	46·1	40·5	40·5	37·6	46·6	45·5
TiO_2	0·29	2·90	2·26	1·46	3·35	11·8	10·5	12·1	3·31	4·1
Al_2O_3	7·64	8·59	8·48	9·30	9·95	8·7	10·4	8·74	12·5	13·9
FeO	19·7	21·0	22·5	18·6	20·7	19·0	18·5	21·5	18·0	17·8
MnO	0·21	0·28	0·29	0·27	0·28	0·25	0·28	0·22	0·27	0·26
MgO	16·6	11·6	11·2	9·46	8·1	7·6	7·0	8·21	6·71	5·95
CaO	8·72	9·42	9·45	10·8	10·9	10·2	11·6	10·3	11·82	12·0
Na_2O	0·12	0·23	0·24	0·26	0·26	0·50	0·41	0·39	0·66	0·63
K_2O	0·02	0·064	0·03	0·03	0·071	0·29	0·096	0·08	0·07	0·21
P_2O_5	—	0·07	0·06	0·03	0·08	0·18	0·11	0·05	0·14	0·15
S	—	0·06	0·05	0·03	0·07	—	—	0·15	0·06	—
Cr_2O_3	0·41	0·55	0·70	0·66	0·46	0·37	0·25	0·42	0·37	—
Total	99·4	99·77	99·46	99·08	100·23	99·67	99·85	99·58	100·2	100·42

The high-titanium mare basalts (Apollo 11 and 17) are nearly free of phenocrysts and form tight compositional groupings. These basalts are considered to have been derived from shallow source depths (at least for Apollo 17).

Ages of lunar basalts constitute an interesting subject. A great many materials have been dated from the Apollo sites, using many different techniques. Taylor (1975), in order to make valid comparisons of the date of surface formation from site to site, compiled a list of dates for the mare basalts (modified here as Table 4.2) in which most of the dates were obtained by the same investigators, using the same techniques. From these, we can see that marial volcanism dated so far ranges from 3160 to 3960 million years—a span of 800 million years! Thus, when some investigators refer to the 'short' span of lunar volcanism, they mean in comparison to the age of the Solar System, for anyone would agree that a length of time greater than the entire Phanerozoic Era on Earth is not 'short'.

The average density of the mare basalts is about $3350 \, \text{kg m}^{-3}$ which is close to the bulk density of the Moon ($3340 \, \text{kg m}^{-3}$). At the estimated pressures and temperature of the lunar interior, the lunar basalts would transform to a mineral assemblage whose density would be about $3700 \, \text{kg m}^{-3}$ (Ringwood and Essene, 1970). Therefore, the Moon cannot be basaltic throughout.

One of the most important parameters governing the character of lava flows during their emplacement and the ultimate morphology of the landforms of solidified lavas is the viscosity at the time of eruption. Viscosity is dependent primarily upon chemical composition, amount and state of volatiles, and temperature. Murase and McBirney (1970 a) have estimated the viscosity of the Apollo 11 basalts based on measurements of synthetic mixtures having the same composition, although an obvious problem is the inability to include the volatiles. A typical value for the Apollo 11 basalt viscosity is 10 poise ($= 1$ Pa s) at 1400 °C, which is comparable to heavy motor oil at room temperature and is an order of magnitude more fluid than terrestrial basalts.

The low viscosity of the lunar basalts explains some of the seeming anomalies and problems of lunar volcanism. On Earth, no basalt plateau is equal in size to the larger lunar mare units. In theory at least, the size of any given lava flow is limited only by the volume of available magma, its ability to flow, and the terrain. The measured gradients on the lunar mare surface (albeit very small!) are sufficient to account for the observed extent of basalt. Another aspect of the ability of lava to flow is the heat retention problem. Obviously, even though a lava may initially be extremely fluid, if it cools rapidly, then the viscosity will increase rapidly. To assess this properly, Murase and McBirney (1970 b) measured the *thermal conductivity* of synthetic lunar basalts and found them to be very low, thus indicating the high heat retention of the lunar lava flows. Thus, the presence of extensive lunar mare basalt flows can in part be explained by their chemical and physical characteristics.

4.3.   *Lava flows in Mare Imbrium*

The recognition of flow fronts in Mare Imbrium by Strom (1965) was a step forward in understanding the origin of the maria. Careful mapping of the flow fronts demonstrated that there were a number of individual flow units each of which had travelled 100 km or more from its source. Orbiter pictures, and more importantly, pictures taken by the Apollo astronauts (fig. 4.4) provide excellent material for photogeological study of these flows.

Schaber (1973) considers that the Imbrium flows were erupted in at least three major episodes (fig. 4.5) and on the basis of crater counts on the flow surfaces, suggests that these episodes have dates of 3000 million years (phase I), 2700 million years (phase II) and 2500 million years (phase III).

The lavas appear to have been erupted from near the outer ring of Imbrium, north of the Carpathian Mountains and flowed towards the middle of the basin. The earliest lavas flowed into the lowest area of the Imbrium Basin near Sinus Iridum and developed extensive pools. The source area of these early-phase lavas is covered by the later flows, but if it is assumed that they originated from the same source area as the later lavas, then they travelled the astounding distance of some 1200 km from their source. Even the later lavas travelled great distances (about 600 and 400 km respectively for phases II and III) on slopes with gradients between 1 in 100 and 1 in 1000. The total volume of these identifiable individual flows in Imbrium is on the order of 4×10^4 km^3 erupted probably over a period of at most about 500 million years.

Lava flow fronts can be measured by the lengths of shadow that they cast and it is estimated that they range from 10 to 65 m with typical heights of 30–35 m. The surfaces of the flows have characteristic basaltic flow features such as well-developed lava channels bounded by levees and also sinuous rilles interpreted to be collapsed lava tubes (fig. 4.2).

There is no major volcanic edifice in the source area for the flows. The source area is identified only by the way the flows narrow and converge backwards on one area and beyond this area no flow-fronts are seen. Centres of volcanic eruption are often strikingly obvious on

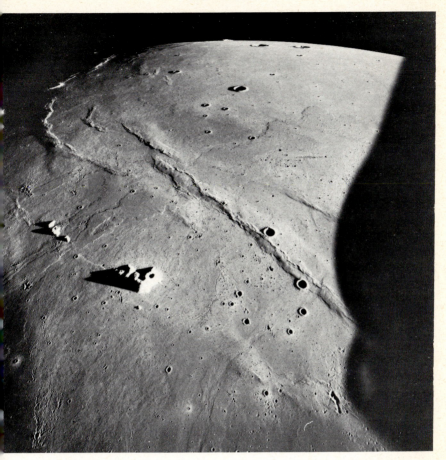

Fig. 4.4.   Oblique view northward across Mare Imbrium showing the basalt flows of
Eratosthenian age and typical mare ridges. The prominent feature in the fore-
ground is Mons la Hire, a mountain block protruding through the lavas. (NASA
Apollo 15–1555.)

Earth by the presence of large central volcanoes and volcanic cones,
formed because lava or pyroclastic material has piled up around the
vent area. For flood basalts, however, the lava is erupted from fissures
at a high rate and does not pile up near the vent, but travels away from
it for considerable distances, leaving little evidence of the original
source. However, careful examination of the Apollo pictures shows
lines of small cones that may mark the fissures from which some of these
lavas erupted. On the basis of such careful studies Schaber has concluded that the Imbrium flows erupted from a fissure some 20 km long.

Studies of differences in colour in the maria by Whitaker (1972)

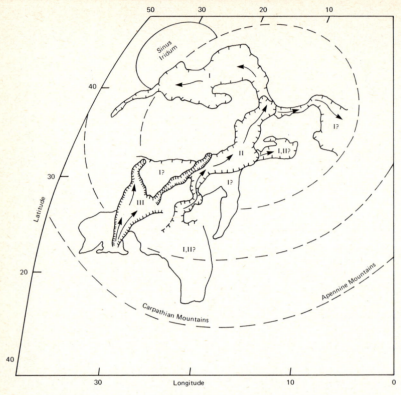

Fig. 4.5.   Map of distribution of Imbrium lavas. Lavas of phases I, II and III are indicated, phase I being the oldest. (From Schaber, 1973.)

show that in Imbrium slight colour differences correspond to individual flow units and that the younger flows tend to be bluer while old flows are redder (fig. 4.6). It is not entirely clear what these colour differences mean; although they may reflect a process of ageing on the lava flow surfaces, it is more likely that they represent differences in composition. Certainly Apollo samples that have high titanium contents are in relatively blue areas while in redder areas the rocks have a titanium content more typical of terrestrial lavas. Thus it might be that as eruptions continued in Imbrium the lavas became richer in titanium.

4.4.   Other maria

Although in Imbrium the individual flows and their surface characteristics can be seen clearly, this is not typical of the maria in general where even at relatively high resolutions flow-fronts are rarely seen. The Imbrium flows have relatively high fronts making them easy to pick out, but careful crater counts on mare surfaces elsewhere show distinctly different crater populations which may correspond to different ages for different parts of the maria. This suggests that some maria are made up of many different sheets superimposed on one another and interdigitating rather than whole mare surfaces representing one single lava sheet. However, these mare surfaces are older than those in Imbrium and if they consisted of thin flows, say just a few metres thick, because of their low viscosity, then they would not be easily recognizable. In addition, the low flow fronts would tend to be masked by regolith. Based on crater counts the eastern maria appear to be older than those to the west.

Holcomb (1971) has made the interesting observation that in some

BLUE INTERMEDIATE RED

Fig. 4.6. Diagram of nearside of Moon showing major colour units of mare plains derived from Earth-based telescopes. (From Head, 1976, after Whitaker *et al.*, 1971.)

mare areas (notably near Lubiniezky, fig. 4.7), where lava has been ponded, the mare area is surrounded by mare terraces standing some 20–60 m above the main mare surface. He considers these to indicate that the lava surface once stood at this higher level but that, following the formation of a surface crust, lava drained from the basin causing a lowering of the surface and leaving a higher terrace round the edge of the depression where lava was congealed and stuck to the surrounding hills.

Howard *et al.* (1972) have argued that this mechanism operated at the Apollo 15 site near Hadley Rille. Mare material is partly enclosed in a structural valley nearly 100 km long and 15 km wide between two mountain ranges at the margin of Mare Imbrium (see figs. 4.1 and 4.8). The mare material joins the main Imbrium mare surface through a 10 km wide gap in the mountain. Surrounding the valley the astronauts observed lava terraces 80–90 m high and interpreted them as 'high lava' marks. This may provide an explanation of the sinuous Hadley rille that runs through the valley, Howard *et al.* suggesting that on eruption the lavas ponded in the valley before they broke through the gap to flow out of the depression. Drainage out of the valley would have lowering of the surface and leaving a higher terrace round the edge of drainage conduit (fig. 4.9).

4.5.   *Sinuous rilles*

A further clue to the style of volcanism in the maria is given by sinuous rilles (figs. 1.6 and 4.2). These interesting features consist of winding channels that usually have a rimless pit at one end. They appear to originate in the pit crater and run down-slope for many tens or even hundreds of kilometres, tending to become narrower lower down. Although the origin has caused considerable controversy in the past and various models for their origin have been put forward, it now appears that they are directly related to the eruption of the marial lavas in which they are found. Two explanations for this have been put forward: one is that they were formed as lava channels and lava tubes, and the other is that they were eroded by rivers of flowing lava.

The mechanism of lava tube formation on Earth has been discussed by Greeley (1971). Lava tubes form only in fluid lavas with low viscosities. When fluid lava spreads out away from the vent the top surface begins to form a crust owing to surface cooling. This crust will thicken in the areas where the flow is moving more slowly and it is only in the regions of more rapid movement that the flow remains fully mobile restricting flow to well-defined channels in the lava. This channel may itself form a crust, leaving a subterranean river of lava confined to a winding pipe bounded by congealed lava. Where the flow of lava is rapid in a narrow channel, surface congealing may be insignificant, but banks (known as levees) can build up along the flanks of the flow by over-spilling of the lava or the accretion of spatter (clots of pasty lava). These banks tend to arch over the channel to merge in the middle, forming a roof and turning the open channel into a closed tube. Although tubes are often crusted over for the whole of their length, it is

Fig. 4.7. Mare material filling depressions on the lunar farside. The distinct benches (arrow) on the margins of the mare unit are considered to have formed by subsidence of the mare lavas by drainage of lava out of the confining depressions through lava tubes and channels. (NASA Apollo 15 2358.)

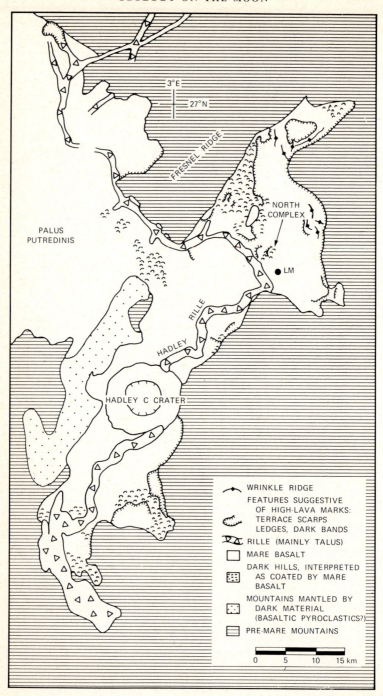

PALUS
PUTREDINIS

3°E
27°N

FRESNEL RIDGE

NORTH
COMPLEX

● LM

RILLE

HADLEY

HADLEY C CRATER

WRINKLE RIDGE

FEATURES SUGGESTIVE
OF HIGH-LAVA MARKS:
TERRACE SCARPS
LEDGES, DARK BANDS

RILLE (MAINLY TALUS)

MARE BASALT

DARK HILLS, INTERPRETED
AS COATED BY MARE
BASALT

MOUNTAINS MANTLED BY
DARK MATERIAL
(BASALTIC PYROCLASTICS?)

PRE-MARE MOUNTAINS

0   5   10   15 km

Fig. 4.9. View of Hadley Rille taken by astronauts near the Apollo 15 landing site. (NASA Apollo 15–85–11421.)

(*Opposite*)

Fig. 4.8. Geological sketch map of Hadley Rille near the Apollo 15 landing site (LM). (After Howard *et al.*, 1972.)

common for there to be windows, or 'skylights' in the roof of the tube, providing a view of the flowing lava beneath. Once a flow has stopped, if the lava is still fluid enough it will flow out leaving a hollow tube within the lava. With time the thin roof may collapse giving once again an open channel. Since lavas flow under gravity, they tend to use existing topographic lows such as old valleys and graben troughs. It is along the axes of such topographic lows that the lava flow will be at its thickest and this is the most likely place for lava tubes and channels to develop; thus an existing valley or structural feature, even when it is completely flooded by later lavas may still have its axis marked by a lava channel or tube. Channels may develop in more viscous lavas but they normally do not crust over to give lava tubes, and the presence of lava tubes is a morphological indication of low viscosity.

Lunar sinuous rilles have the characteristics of unroofed terrestrial lava tubes. However, while terrestrial lava tubes tend to be only a few metres or tens of metres across, sinuous rilles may be considerably bigger than this, their depths in some cases being up to 300 metres. The much lower gravity on the surface of the Moon would allow lava bridges to be wider than on Earth and the calculated maximum width for a lunar lava bridge corresponds well with the width of sinuous rilles (Oberbeck *et al.*, 1969). It is therefore argued that although some sinuous rilles may be lava channels, over much of their length they are lava tubes whose roofs have been destroyed by meteoritic impacts and moonquakes. Quite frequently sinuous rilles terminate without any narrowing of the channel and re-appear some kilometres further down slope (fig. 4.16). This is taken as evidence of the presence of an uncollapsed roof still existing over the tube.

The lava erosion hypothesis is less well documented from terrestrial examples. Hulme (1973) has argued that although some sinuous rilles may be lava tubes or channels, many were produced by the erosion of underlying lavas by a fast-flowing turbulent lava for a long duration. He argues that the highly fluid lava erupted on the lunar surface would form a narrow stream. The middle of the stream would be turbulent and flowing rapidly, and its sides would tend to spread very slowly and would soon congeal. He argues that the stream width would then be fixed and would be essentially uniform along the whole length of the flow. Lava in the middle of the stream would erode a channel for itself by melting the underlying rocks. Although some widening of the channel would occur by erosion, the main effect would be downward, cutting a steep-walled channel. With distance away from the lava source the eroding power would decrease until it reached zero; the lava would continue to flow but would not be hot enough to continue eroding and the lava would be emplaced as a pond or thin sheet. Channels could be the result of several eruptions following the same paths.

Carr (1974) has also considered the role of lava erosion in the formation of lunar rilles. Unlike Hulme, he considers that 'erosion will occur once a tube or channel is established and that the large dimensions of many lunar rilles are attributable to this erosive action'. Greeley and Hyde (1971) have found terrestrial evidence of lava erosion of both lava and country rock. Peterson and Swanson (1974) from their observations of active flows in Hawaii have argued that sustained high temperatures within the lava tubes allow flowing lava in the tube to erode the floor. If this is true for terrestrial tubes then it is almost certainly true for the Moon. The amount of lava erosion that has taken place in lunar tubes and channels is not certain as this depends on the length of time that fluid lava was flowing within the channel. However, Carr calculates that a reasonable erosion rate would be about 1 m/ month. Using this figure many of the large sinuous rilles that Carr attempts to explain in terms of erosion in channels would require sustained eruptions of periods of several years. We consider that channels and tubes will probably be large because of the style of eruption, but erosion during continued flow will almost certainly make them larger than they might otherwise have been.

Whatever emphasis might be placed on either of these mechanisms, it is clear that sinuous rilles mark the paths of marial flows, and their distribution in the maria gives some idea of the directions taken by lava during the emplacement of the maria.

The origin of the highland sinuous rilles may be relevant to the geology of the maria. Highland sinuous rilles are much wider than those of the maria, are less sinuous, and the larger ones are floored by mare material which has a highly sinuous mare-type rille within it. The origin of these highland rilles is not clear; but since they are not within recognizable lava flows it is unlikely that they initially formed as lava tubes or channels. However, like the rilles in the maria, they have rimless craters at their heads and are clearly related to volcanic activity because of the presence of mare material which appears to have flowed from the craters at the head of the rilles. John Murray has argued that rilles of this type result from erosion of unconsolidated highland materials by the flow of lava. While this mechanism may explain some features of the highland sinuous rilles, Schröters Valley (a unique highland rille) was originally a closed depression (fig. 4.10) and the mare material broke over the side of the valley before it reached the valley end. Some other explanation is required for this valley and it

Fig. 4.10. Oblique view southward across the Aristarchus Plateau showing Schröters
Valley, an enormous 'highland' rille more than 150 km long. It consists of a wide
outer depression inside which is a normal mare-type sinuous rille which continues
off the plateau onto the adjacent mare surface. The fresh crater at top left is
Aristarchus. (NASA Apollo 15–2611.)

may be the result of volcano–tectonic activity and subsequent collapse.
Although these valleys need more study, they suggest that some of the
marial lavas originated from sources outside the maria in the highland
regions and flowed down through the highlands to spread out on the
mare surfaces.

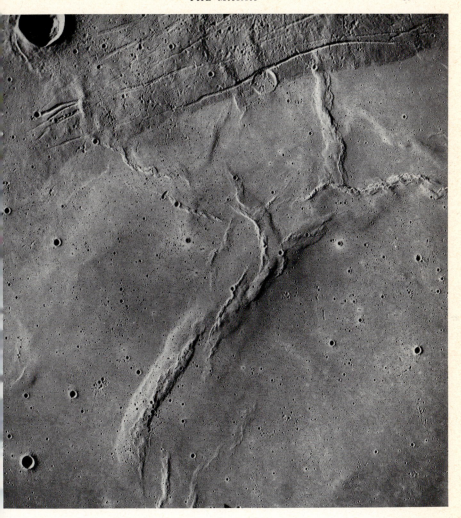

Fig. 4.11. Mare ridges and graben (straight and curved rilles) on the edge of Mare Serenitatis. Under these low lighting conditions the broad mare ridges surmounted by steep-sided narrow ridges are clearly seen. The graben cut darker mare material that is older than the light mare material in middle and lower part of picture. (NASA Apollo 17–0451.)

4.6. *Mare ridges (Wrinkle ridges)*

These are shallow ridges or ridge-complexes that can be hundreds of kilometres long (fig. 4.11). They are generally found in mare regions, but occasionally extend onto the adjacent highlands. Some authors, particularly in the earlier literature, have called them wrinkle ridges.

Individual ridges may be up to several kilometres wide and have heights of several tens of metres. Strom (1971) recognized, from a careful study of Lunar Orbiter pictures and Earth-based telescopic pictures, that there were two types of closely associated ridges. The first of these are low, broad ridges only seen at low sun-angles, while the second type are usually more prominent because they are steeper, have a 'ropy' appearance and often occur on the crests of the broad shallow ridges.

There have been many ideas about the origin of these features. Strom argues that the broad ridges are up-arched surface strata produced by magma in fissures below the surface pushing up, while the narrow, steep-sided ridges are extrusions of this same magma from cracks in the crest of the anticline. It is noticeable that some of the ridges are asymmetrical in cross-section (fig. 4.12) and have one side steeper than the other; this together with other evidence has suggested to Howard and Muehlberger (1973) that they represent thrust faulting.

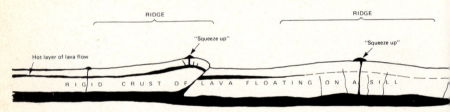

Fig. 4.12. Hypothetical cross-section through mare ridges of two different types. On the right is 'normal' ridge; on the left is an asymmetrical ridge.

Although there is no close agreement on the nature of these structures, it is generally agreed that they represent folding and faulting of the surface layers of the maria—but are they deep-seated or shallow structures? If they are deep-seated, they must be surface expressions of tectonism in the interior of the Moon. Alternatively they may be shallow structures formed by local movements within the maria. Hodges (1974) has proposed that they are analogues of features that typically develop on surfaces of lava lakes, such as those in Hawaii. Lava lake ridges form on a congealed surface layer over liquid lava where the liquid lava has moved by convection, dragging the crust into ridges and over-riding slabs. As Hodges herself points out, the weakness of this model is that on the Moon mare ridges are often continuous across lavas of different ages and sometimes extend for short distances into the surrounding highlands; also there are several orders of magnitude difference in scale. However, if we consider a slightly different model (Guest and Murray, 1976), in which the maria consist of a composite pile of lavas, some of which still have hot plastic interiors, and intruded into this complex are sill-like sheets of basaltic magma below a rigid crusted-over surface, then it would be possible to form the mare ridge-complexes by movement of this surface as a series of rigid plates over the still liquid or plastic magma. These movements would be driven by shallow convection, intrusion, withdrawal of magma and sagging of the central portion of the basin; individual plates on the surface would move relative to one another giving buckling, low-angle thrusting, squeeze-ups, tension cracks and associated auto-intrusion, and also local strike slip ridges. This mechanism would explain all the features of mare ridges including the way that ridges tend to parallel adjacent topography and form over buried craters.

4.7. Calderas and individual vents

As well as the peppering of circular impact craters that cover most of the mare surfaces, there are other craters which either lack rims or have very shallow rims (fig. 4.13). This morphology implies that these craters formed principally by collapse rather than ejection of material which would have built up a rim. These craters strongly resemble the typical caldera of the Kilauean type found in basaltic volcanoes on Earth. Calderas usually form by collapse when large volumes of volcanic material are erupted or when magma withdraws from a shallow magma reservoir allowing surface collapse. The lunar calderas are considered to be counterparts of terrestrial collapse calderas such as those in Hawaii, and the large size of some of them (up to several kilometres in diameter) is to be expected with the large scale basaltic eruptions that we know took place. However, many of them are relatively small, being about one kilometre in diameter.

Other individual vent features in the maria include low domes, steep domes, pyroclastic cones and steep-sided lava flows. The low domes have smooth convex slopes of just a few degrees typically rising to a summit crater (fig. 4.14). There are about 80 of these on the nearside of the Moon and many of these occur in clusters; for example, about half of them are found in the Marius Hills which is one of the largest centres of igneous activity on the Moon. Most of these shield volcanoes are about 10 km across, which is equivalent to small shields on Earth. Summit craters are generally between 2 and 3 km across. Shields may be inferred to result from volcanism localized on a central conduit and their shallow form suggests that they were formed largely from quiet effusion of fluid lava with little pyroclastic activity. By contrast with flood lavas, those of the shields were probably relatively short (up to

Fig. 4.13.   Possible calderas (each about 5 km across) in mare materials near the Marius Hills, Oceanus Procellarum. Note irregular shape and slight raised rim of caldera depressions. (NASA Lunar Orbiter IV 150 H$_2$.)

about 5 km) and were built up by successive eruptions on a single vent. Eruption rates were probably lower than with the flood lavas.

Steep domes, by contrast, probably formed by the extrusion of relatively viscous lava which piled up around the vent. These domes tend to have rugged upper surfaces and steep slopes of up to 20°. Many of the domes are pitted with rimless craters and have coarse ridges which may be interpreted as the result of flow. Steep domes are common in the Marius Hills (fig. 4.15), and those near Gruithuisen are also of this type. The diameter of a steep dome is usually several kilometres. Although they may have formed from viscous (low temperature) basaltic lavas, it is possible that they are silicic lava produced by differentiation in a magma chamber.

Fig. 4.14. Low domes (each about 5 km across) with summit craters and rifts, in mare material in Orientale. (NASA Lunar Orbiter IV 187 H_3.)

The steep-sided lavas (fig. 4.16) which have lengths up to 6 km but in one case up to 20 km, may also be the result of eruption of lava of a different composition from basalts. The fronts of these lunar flows are lobate with heights of up to 80–100 metres and they normally have low ridges on the flow-surface similar to those on terrestrial flows. At the source of the flow there are commonly pyroclastic cones. The average volume of these lavas is about 4 km^3 which is similar to typical volumes of intermediate (e.g. andesite) composition flows on Earth. The total volume of material of this type presently exposed on the Moon's surface is in the order of 100 km^3 which is extremely small by comparison with the large volume of flood basalts.

Pyroclastic cones are recognized by their wide rims and the elevation of the crater, the floor of which is above the surrounding terrain; these characteristics distinguish them from typical impact craters. Also the volume of the rim material is clearly greater than the volume of the

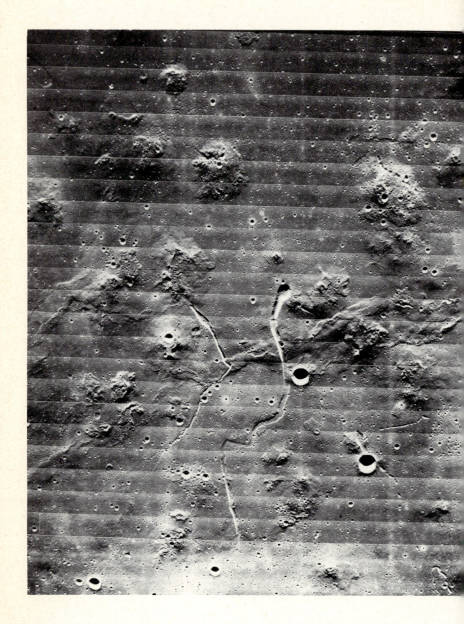

Fig. 4.15.   Low and steep domes and steep sided flows in the Marius Hills. Two sinuous
rilles (middle of picture) cut a mare ridge complex. Partially roofed rilles (lava
tubes) occur in lower right. North is to the left. (NASA Lunar Orbiter V 213 M.)

Fig. 4.16. A thick short lava flow (arrows indicate direction of flow). The flow originates in a cone (P) which has several breached craters. A steep-sided mare ridge (R) appears to have flowed into a crater (F). (NASA Lunar Orbiter V 216 H₃.)

crater itself. These cones are typical of the Marius Hills area, but they occur elsewhere, sometimes as aligned clusters on fissures. Some aligned features may be spatter cones (fig. 4.17). Although the marial lavas appear to be deficient in water, at least in the samples returned to Earth, it is clear that gases of some kind were emitted at volcanic vents in localized areas on the Moon. These explosive vents are often associated with the more viscous magmas, as is observed on Earth.

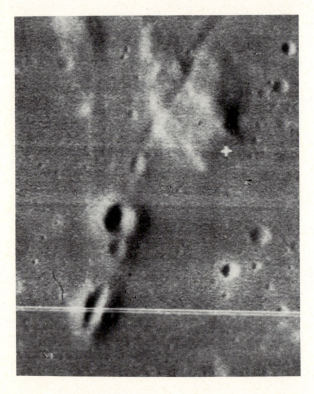

Fig. 4.17.   Small (about 1 km across) presumed pyroclastic cones on graben in Oceanus Procellarum. (NASA Lunar Orbiter photograph.)

Apart from a few features described as pyroclastic cones, there are few pyroclastic landforms in the maria, considering the vast extent of basaltic plains. This paucity might be explained in several ways. First, lunar cinder cones may be rather small features which, after 3000 million years or more of impact erosion, are no longer recognizable. Second, and more importantly, McGetchin and Head (1973) have shown that, under the reduced gravitation field, pyroclastic particles would travel further on the Moon than on the Earth, with the result that pyroclastic deposits would tend to spread wider, giving lower profile cones.

Glasses from the lunar soils are evidence of lunar pyroclastic activity. Several types of glass bead have been recovered from soil samples. Although much of the bead material can be attributed to droplets of impact melt, some is very probably of volcanic origin—notably the emerald green glass of Apollo 15 and the orange glasses of Apollo 17 and elsewhere as summarized by Taylor (1975). These glasses are of uniform size and composition, contain high concentrations of volatiles and were formed contemporaneously with the marial basalts with which they are associated. It has been suggested that these types of glasses represent fire-fountaining (literally fountains of red-hot lava), perhaps analogous to that observed during eruptions in Hawaii.

4.8. *Mechanism of flood basalt eruption*

How can the marial basaltic volcanism be explained? Flood basalt fields on Earth generally consist of thick piles of multiple flows that extend for tens of kilometres. In certain areas such as the Columbia River Plateau, flows may be as thick as some tens of metres and several hundred kilometres long, comparable in size to some of the marial basalts. Why were the lunar lavas so extensive? There are many factors that may control the length of a lava flow. For example, low viscosity lavas will tend to flow further than viscous ones; lavas erupted on steeply sloping surfaces will flow further than on almost flat surfaces; and flows constricted to narrow valleys will become thicker, thus increasing their hydrostatic head, enabling them to flow further. Although in the early stages of basin filling by marial material the underlying surfaces may have had topographic relief, much of this relief would have been filled in rapidly by the lavas, giving the fairly level extensive surface which we now see. Thus for the lavas visible to us now, topography probably played little part in governing the length of flow.

The viscosity of the lavas is important. Studies of marial lavas have shown that at their melting temperature (about 1400 °C) they have viscosities of less than 100 poise (10 Pa s) and in the case of the high titanium basalts less than 10 poise. Columbia River basalts, on the other hand, which are some of the most fluid erupted on Earth, have viscosities greater than 100 poise. The low viscosity of the mare lavas would have allowed them to spread at higher velocities than typical terrestrial basalts and so produce extensive flows before viscosity was increased by cooling and the flows slowed down by congealing. Murase and McBirney (1970 a) have estimated that, taking into account the reduced lunar gravity, marial lavas of the experimentally measured viscosities could have travelled twice the maximum distance travelled by terrestrial basaltic lavas.

The length of a lava flow also depends on the hydrostatic head. On an almost horizontal surface, as we have in the younger maria, this will be related to the thickness of the lava, a thicker flow being able to flow greater distances than a thin one. Dănes (1972) has argued that from a theoretical standpoint the length of the lava flow of such a surface is proportional to the cube of the flow's thickness. He considers that on the Moon the flow thickness would have to be less than a factor of three greater than a typical Columbia River basalt lava to give the enormous lengths of lava we see. The thickness of over 100 metres for lunar lavas would satisfy this condition.

The effect of decreasing the viscosity of the lava is to produce thinner flows as their spreading capacity is increased. Thus in order to get a thick flow of low viscosity near the source, as is required to give the hydrostatic head for such long flows, the rate of lava eruption must be high. In fact Walker (1972) has argued that it is rate of eruption which primarily controls the lengths of lava flows. Lavas of low viscosity, if erupted slowly, will produce a number of thin relatively short flows superimposed on one another, producing a pile of lavas around the vents. With increased eruption rate these lavas will extend much further from the source.

If this argument is correct, then we can imagine the maria to have been made up of a number of very rapid eruptions of lava. However, it is likely that these individual eruptions were separated by long intervals of time. The area of the maria is about 17×10^6 km^2. If we assume an average thickness of about 2 km for the lava we arrive at a volume of about 34×10^6 km^3 and if we assume that this was erupted over approximately one billion years this would give an *average* output of around 1 cubic metre per second. Even if this estimate of the volume is wrong by a factor of 2 or 3 the output is only a few cubic metres per second for the whole of the Moon and this is comparable to output from *individual* volcanoes such as Etna or Hawaii. Therefore the *total* eruption rate for the Moon is very low by terrestrial standards. But for *individual* eruptions, the eruption rate was high, implying long periods of repose of as much as several thousand to tens of thousands of years between each eruption.

The styles of basaltic volcanism that occurred on the Moon are reflected in the preferred surface features as discussed by Greeley (1976). Basaltic landforms are the end result of many complex factors, including viscosity (itself the result of lava composition, temperature, volatile content, and others), rate of eruption, and topography. Considering only effusive activity of fluid lava, three types of surface can be distinguished on Earth: (1) *Flood basalts*—characterized by high eruptive rates from fissures producing massive, often ponded lavas that lack extensive flow surface features, (2) *Shield-forming basalts*—produced by lower rates of eruptions than flood basalts and typified by the 'leaky' magma reservoirs at Hawaii; here eruptions are localized round a central conduit and result in thin flows with abundant flow features such as lava tubes and channels developed during prolonged but sporadic activity, and (3) *Plains-forming basalts*—an informal name applied to a type of basaltic volcanism that is intermediate between flood basalt activity and shield-forming volcanism. Plains-forming eruption, typified by the Snake River Plain, Idaho, involves rates of eruption comparable to those of Hawaii from fissures, or from point sources along rift zones. The result is a gently undulating plain composed of multiple very low-profile shields (slopes typically less than $1°$) that overlap one another, and interspersed, long tube-fed flows. Lava tubes and channels are common both on the gentle flanks of the small shields and intervening flows.

Mare surfaces exhibit both flood basalts and plains-type surfaces; massive constructional shields of the Mauna Loa variety are lacking, although small shields occur in several areas. From limited Apollo and Luna sampling sites, there is the suggestion of a correlation between volcanic style, and age and composition of mare basalts (table 4.2). Mare Serenitatis (Apollo 17), Mare Tranquillitatis (Apollo 11), and Mare Fecunditatis (Luna 16) are generally older mare surfaces composed of high Fe and Ti composition basalts, while part of Oceanus Procellarium (Apollo 12) and Mare Imbrium (Apollo 15) are younger basalts of slightly more silicic lavas. The older mare surfaces seem to have a general lack of flow features, e.g. sinuous rilles, while the younger surfaces contain most of the sinuous rilles and thin flows observed on the Moon. These relations suggest a multiple emplacement history, in which initial eruptions were of the flood basalt variety, involving relatively high rates of effusion, followed by lower rates of eruption of slightly more viscous lavas to produce sinuous rilles and other thin, plains-type lava flows. Observed differences in the distribution of sinuous rilles among the mare basins suggest that the second style of volcanism—plains-forming basalts—did not occur in some basins, such as Crisium and Smythii.

Most volcanoes on Earth are related to structural belts between adjacent rigid crustal plates, known as mobile belts. There does not appear to be this kind of control in the lunar maria and most of the

Table 4.2.   Ages of marial lavas derived from isotopic dating of Apollo samples. (From
Taylor, 1975.)

	Age in million years	Rock type	Dating method
Apollo 14	3960	Al-basalt	Rb–Sr
	3950	Al-basalt	$^{40}Ar–^{39}Ar$
	3950	Al-basalt	Rb–Sr
Apollo 17	3830	High Ti	Rb–Sr
	3820	High Ti	Rb–Sr
	3760	High Ti	$^{40}Ar–^{39}Ar$
	3740	High Ti	$^{40}Ar–^{39}Ar$
Apollo 11	3820	Low K	$^{40}Ar–^{39}Ar$
	3710	Low K	Rb–Sr
	3630	Low K	Rb–Sr
Apollo 11	3680	High K	Rb–Sr
	3630	High K	Rb–Sr
	3610	High K	Rb–Sr
	3590	High K	Rb–Sr
	3560	High K	$^{40}Ar–^{39}Ar$
Luna 16	3450	Al-basalt	$^{40}Ar–^{39}Ar$
	3420	Al-basalt	Rb–Sr
Apollo 15	3440	Quartz basalt	Rb–Sr
	3400	Quartz basalt	Rb–Sr
	3350	Quartz basalt	Rb–Sr
	3330	Quartz basalt	Rb–Sr
	3320	Olivine basalt	Rb–Sr
	3310	Olivine basalt	$^{40}Ar–^{39}Ar$
	3260	Quartz basalt	Rb–Sr
Apollo 12	3360	Olivine basalt	Rb–Sr
	3300	Olivine basalt	Rb–Sr
	3300	Olivine basalt	Rb–Sr
	3270	Quartz basalt	$^{40}Ar–^{39}Ar$
	3260	Quartz basalt	Rb–Sr
	3240	Olivine basalt	$^{40}Ar–^{39}Ar$
	3240	Quartz basalt	$^{40}Ar–^{39}Ar$
	3180	Quartz basalt	Rb–Sr
	3160	Quartz basalt	Rb–Sr

structural control appears to have been provided by deep-seated
structures generated by the large impact basins. Centres of volcanic
activity are more common in regions surrounding the basins and also
in areas between two adjacent basins where dislocation of the crust
may have been intense. However, the lack of marial volcanism on the
far side of the Moon is interesting and may possibly be explained by
the lunar asymmetry (the centre of figure of the Moon is displaced from
the centre of mass). The crust on the farside may be thicker than on the
nearside and thus fractures induced by basin impact may not have
penetrated to the asthenosphere* on the farside.

*On Earth the brittle layers at the surface are known as the lithosphere and the plastic
layers below as the asthenosphere.

5. Impact cratering mechanics

5.1. *Introduction*

The origin of lunar craters has been strongly debated, but in recent years the impact mechanism has gained almost total acceptance. Craters can form either by the ejection of the material or by removal of material from below causing the surface to collapse. Ejection craters usually have raised rims consisting in part of material thrown out from the crater and include volcanic explosion craters, chemical and nuclear explosion craters and impact craters. Collapse craters on the other hand do not usually have raised rims and include subsidence pits, volcanic pit craters and calderas, some of which may also have raised rims if ejection of material accompanied collapse.

The best preserved terrestrial impact crater is Meteor Crater, Arizona (fig. 5.1). Until the 1950's geologists were reluctant to accept an impact origin for this crater, preferring to resort to the more familiar volcanic processes to explain it. Daniel M. Barringer, who acquired the crater in 1903, was convinced of its impact origin, but his idea was not properly accepted among geologists until the classic work of Shoemaker (1960). His study showed that unlike volcanic craters the rim of Meteor Crater consists of structurally uplifted bedrock overlain by an overturned flap of the country rocks in which the original stratigraphy is preserved, but in the reverse order, the oldest strata below the crater occurring at the top of the sequence on the rim. Shoemaker showed that these structures were similar to those of nuclear and chemical explosion craters and suggested that the mechanism of crater excavation has many similarities to those of impact.

5.2. *Chemical and nuclear explosion craters*

The term 'explosion' can include two different types of cratering processes. One of these, the classical explosion by definition, is produced for example by the deflagration of gunpowder and consists of combustion giving rapid expansion of gases. The pressures generated depend on the thermal expansion of the hot gases and the extent to which they are confined by the surrounding medium. Volcanic

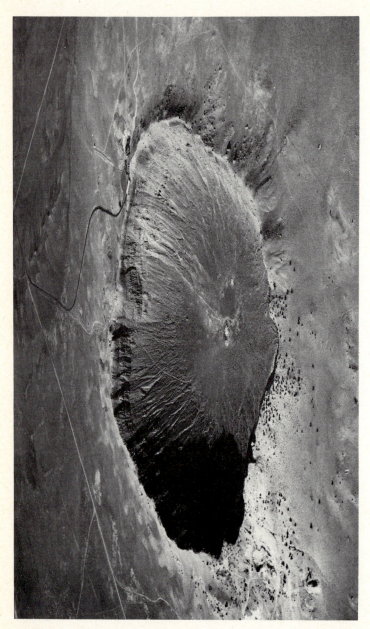

Fig. 5.1. Oblique view of Meteor Crater, Arizona. Note large blocks of rock on rim. The museum building on the far side gives the scale. (Photograph: D. Roddy, U.S. Geological Survey, Flagstaff, 1976.)

explosions are similar in character but operate by adiabatic expansion of gases such as the flashing of water into steam. Pressures in processes of this type generally do not exceed five kilobar.

Craters produced by high-energy chemical explosions, for example when TNT is detonated, or by nuclear explosions, are excavated by a shock-wave process rather than by simple gas expansion and are quite different from volcanic explosion craters. The shock-wave is a pulse of instantaneous compression involving ultra-high pressures propagating away from the source, the shock-front advancing at supersonic velocity in excess of several km per second. When the shock-front interacts with the ground surface, a region of decompression is generated and this expands back into the compressed medium as a *rarefaction phase*. The crater is formed behind the decompressing region as material is accelerated towards the free surface and ejected above the ground. During the very brief period of the shock pulse, pressures in the range of hundreds of kilobars up to megabars are generated and rocks may be intensely altered, or even melted or vaporized by transient high pressures and temperatures in nuclear explosions. This alteration of rocks is known as shock-metamorphism.

As well as cratering produced by the rarefaction or tension waves, there is a second mechanism that operates in buried high explosive and nuclear explosions, and to a much lesser extent in surface high energy explosions. This other process is known as gas acceleration in which disaggregated fragments are given an additional thrust from high pressure gases escaping rapidly from the explosion cavity. The gas is initially produced as a large bubble of vaporized material surrounding the nuclear charge; expansion of the gas causes the ground surface below the bubble to swell up and eventually burst, throwing out material at an early stage in the event.

When these high energy explosion craters are examined, their rims are found to be complicated structures consisting of uplifted and overturned strata, ring anticlines and synclines, and thrust and normal faults (fig. 5.2); tangential flow causes great thrust slices to be injected outwards into the rims. With near-surface blasts, which are more closely analogous to impact than those of deeply buried charges, studies of the movement of buried markers show that the rocks below the crater floor tend to move inwards and upwards producing hills of bedrock in the floor of the crater.

Outside the crater is a layer of continuous ejecta blanket which thins away from the crater and becomes discontinuous at the outer edge. Numerous small craters often forming clusters and chains surround the continuous ejecta. These are secondary impact craters

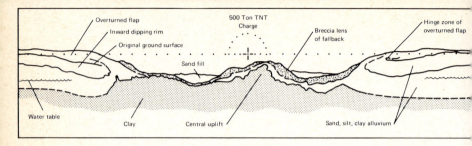

Fig. 5.2. Cross-section of Snowball crater (Canada) a typical high energy artificial explosion crater. The formation of the central uplift may have been assisted by the high water table, the water lubricating the rocks below crater and giving them mobility. (From Roddy *et al.*, 1976.)

formed by impact of individual missiles ejected from the primary event. Much of the ejecta from craters formed by shock-waves hits the ground at high enough velocity to cause cratering. Because much of this material is ejected at high velocity, when it impacts it will excavate considerably more material than its own mass. Ejecta sheets so formed consist of much local bedrock. Thus close to a primary crater (fig. 5.5), the volume of ejecta arriving on the surface is so large that it will swamp individual secondary craters; but with distance, the volume of primary ejecta will decrease and secondary craters become visible.

5.2. *Laboratory impact studies*

Impact events are similar to high explosive and nuclear explosions because in all these events much of the excavation is done by shock-wave processes; but impacts differ in not having a gas acceleration phase as a necessary part of the mechanism.

The understanding of impact cratering events has been considerably advanced by laboratory experiments using a high velocity gun, such as that installed by Donald Gault at NASA Ames Research Center in California. This gun can fire small pellets of various materials into targets contained in a chamber maintained at low pressure to reduce the effect of air drag on the ejected particles. The gun can be rotated to fire projectiles at different angles into the target. Normally these targets consist of unconsolidated sand layers containing a small amount of epoxy resin; after a crater has been formed (fig. 5.3), the target is heated to set the resin and can be cut in half to examine the deformation of the coloured layers (fig. 5.4). Targets simulating rocks with interbedded layers of different physical properties, such as strength, can be produced by spraying the sand surfaces of the target

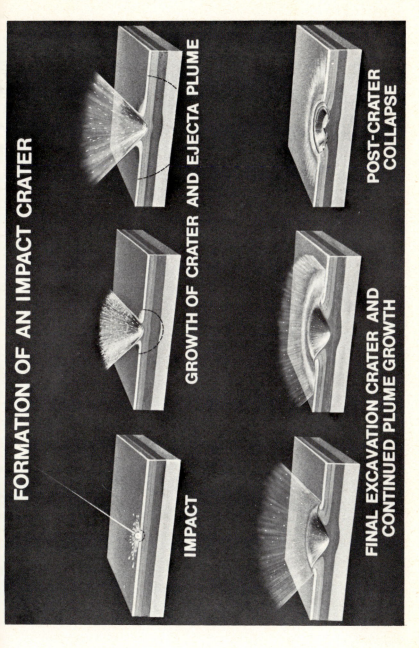

Fig. 5.3. Schematic presentation of ejecta patterns and crater development for a laboratory impact in homogeneous, non-coherent sand. (Courtesy D. Gault.)

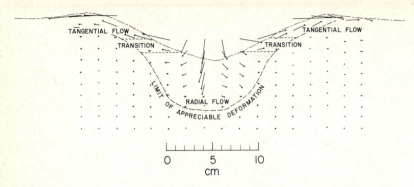

Fig. 5.4.   Flow diagram derived from impact crater experiments at the high velocity
gun, NASA–Ames Research Center. Figure shows total movement produced by
the cratering process in individual unit masses. (From Gault *et al.*, 1968.)

with lacquer before putting on the next layers of sand; joints and
fractures can be simulated by leaving gaps in the lacquer.

In the early stages of impact (at velocities of 5–10 km s^{-1}) when the
target is being compressed, the mechanics differ from those produced by
high explosive or nuclear events. Two shock-fronts are produced at the
instant of impact, one in the projectile and the other in the target
medium. As the projectile penetrates the target, shock-waves run
upwards into the projectile and downwards into the target and at this
early phase, shock-compressed material is limited to a lens-shaped
mass directly below the penetrating projectile. The deformational
stresses are at least 10^3 to 10^4 times greater than the strength of the
target material and the materials flow hydrodynamically.

As the projectile penetrates more deeply into the target, an ever-
increasing mass of target and projectile is engulfed by the shock-waves
which are themselves modified by the presence of free surfaces, not
only at the top of the target, but also on the sides of the projectiles. The
free surfaces cannot sustain this state of stress and, as the shock-waves
race along across these surfaces, a region of rarefaction develops
behind the shock waves to decompress from the high pressure conditions. When the rarefaction waves appear, the process of 'jetting'
starts and then decays rapidly. This is the hydro-dynamic ejection of
mass at very high velocities and both projectile and target are melted
and vaporized. Material is squirted out from near the interface between
the compressed target and projectile at velocities that are several
times the original impact velocity. Sufficient heat to melt the target
can also be generated and the crater may be lined by melt which
remains as a lens-shaped pool in the floor of the final crater. The end of
the compression stage is reached when the shock-wave reflects from

the backside of the projectile and the projectile has been consumed by shock-waves.

The main mass of material is then ejected from the crater at much lower velocities during what is known as the excavation stage. The shock-wave geometry consists of an approximately hemispherical shell of compressed material expanding radially outwards. This distributes the kinetic energy from the projectile over a progressively greater mass of target material. The crater is excavated by material being thrown upwards and outwards at progressively lower velocities. At the last stage of crater excavation when velocities of the excavated rock are low, the rim is formed by uplift and overturning of the crater edge. Excavation ends when the stress decays to a level which equals the strength of the target rock. However, the stress wave continues to expand and decay beyond the limits of the crater rim to become a simple elastic wave which for large impacts could be highly destructive to surface features for some considerable distance away from the crater.

As Gault (1974) has pointed out, for a given energy of impact the ultimate size and shape of an impact crater is determined principally by two parameters, one being the strength of the material excavated and the other the gravitational acceleration at the surface of the planet. The relative significance of these two parameters will depend on the size of the impact being considered. The importance of the material strength becomes less significant with larger craters, but as the crater becomes larger, resistance to gravity will increase.

The ultimate shape of an impact crater depends partly on post-excavation modifications. If we consider a crater developed in water, the crater will disappear shortly after impact, leaving no evidence of its previous appearance because water effectively has zero strength and cannot support the cavity. However, even in rock targets, collapse will still occur as the surface attempts to restore its original shape. In larger craters, landslides from the rim into the crater tend to enlarge the diameter of the crater producing a terraced inner rim. With larger craters also it appears that, where the strength of material becomes less important in the cratering process, there will be a rebound effect causing rocks below the crater to move inwards and upwards, lifting the crater floor and giving a large mound called a central peak on the floor of the crater. This may be a similar process to that described for near-surface explosions.

Theoretical studies show that for the Moon the most probable angle of impact is 45° to the surface. Most impacts thus occur oblique to the surface; however laboratory experiments show that only material thrown out during the jetting phase will have an asymmetrical

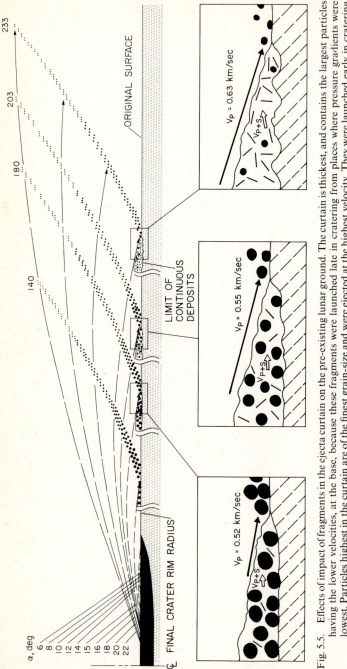

Fig. 5.5. Effects of impact of fragments in the ejecta curtain on the pre-existing lunar ground. The curtain is thickest, and contains the largest particles having the lower velocities, at the base, because these fragments were launched late in cratering from places where pressure gradients were lowest. Particles highest in the curtain are of the finest grain-size and were ejected at the highest velocity. They were launched early in cratering when pressure gradients were highest. The curtain sweeps outward from the crater rim as time increases, depositing finer and finer grained material on the surface at progressively high velocities and angles. Drawings of interaction of the curtain with the lunar surface are shown at the bottom of the figure. Local material (dashed lines) is shown mixed with primary and secondary material being ejected at lower impact velocity than incoming fragments. This gives rise to a ground-hugging flow of mixed primary crater ejecta and local material that flows behind the primary ejecta curtain. (From Oberbeck, 1975.)

distribution around the crater. The bulk of ejecta will be thrown out symmetrically to give a circular crater. Only in the case of rare glancing impacts at a few degrees will the main mass of ejecta be thrown out to give an elliptical crater with an asymmetrical ejecta sheet, most of the material thrown out sideways and down-range rather than up-range.

The experience of modelling impact events in the laboratory together with studies of other artificial shock-wave craters allows a better understanding of the features observed in impact craters. However, it should be remembered that artificial craters are small compared with terrestrial and lunar impact craters and we must extrapolate our knowledge of the basic processes for small craters to those in much larger ones.

5.3. *Examples of impact craters on Earth*

The combination of increasing knowledge of shock-wave cratering from experimental work and the impetus given by the space programme has led to a greater interest in impact craters and a more determined effort to look for terrestrial examples and the criteria for their identi-fication. Thus, while in 1960 only about 16 impact craters were recognized on Earth, eight years later more than fifty had been identified plus another 50 suspected structures (French, 1968).

The Pleistocene Meteor Crater, on the Colorado Plateau in Arizona (fig. 5.1) is a bowl-shaped depression about 1·2 km across and 200 m deep. The country rocks excavated by the impact (fig. 5.6) are nearly flat-lying sedimentary rocks with gentle folds; they consist of three distinct units, the Permian age Coconino Sandstone, the Kaibab Limestone, and the Triassic Moenkopi Sandstone. The surface of the Colorado Plateau in this region is relatively fresh but has been some-what lowered locally by erosion since the crater was formed.

The crater rim stands 50–60 metres above the surrounding plain and consists of bedrock strata that have been overturned to give a reversed stratigraphy in which the oldest strata overlie the youngest. The rim has also been uplifted by 10–14 metres (Roddy *et al.*, 1975) to give an outward dip of the strata; the dip increases upwards in the crater walls as a result of the strata having been 'peeled back from the area of the crater, somewhat like petals of a flower blossoming' (Shoemaker, 1960). These upturned and overturned strata are cut by many nearly vertical faults that strike generally north-east and north-west parallel to regional joint sets in the bedrock. Regional jointing has controlled the shape of the crater which has a somewhat square appearance in plan view, with the diagonals of the square coinciding with the trend of

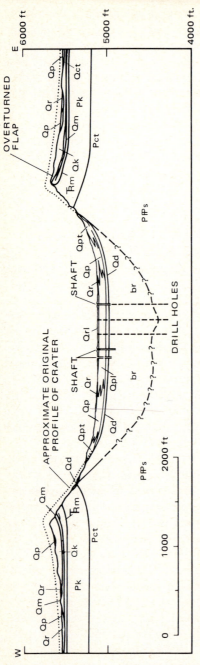

Fig. 5.6.   Cross-section of Meteor Crater Arizona (from Roddy *et al.*, 1975, after Shoemaker).

the principal joint direction. Thrust faults also occur with the sense of movement away from the crater.

By drilling 161 bore holes in the uplifted and overturned ejecta around Meteor Crater, Roddy *et al.* (1975) have determined the stratigraphy of the ejecta and the relations between thickness and distance from the crater. Throughout most of the ejecta, the individual strata are well defined in inverted order, although there is some limited mixing.

Shock metamorphism is represented by frothy glass (lechatelierite) produced by melting of quartz, and coesite, a high pressure form of quartz. Nickel-iron fragments are also associated with the crater.

Because impact processes are well ordered events, rocks from individual strata are not randomly distributed in the surrounding ejecta, but are emplaced systematically. Strata excavated from the deepest level below the crater appear uppermost in the ejecta sequence and near the crater rim (fig. 5.6). This observation has particular value to lunar field geology where samples representing the deepest strata must be looked for at crater rims.

Other terrestrial impact craters are more eroded than Meteor Crater. Brent Crater in Canada (Dence, 1968) is nearly 4 km across and, although its rim has been removed by erosion, it is clearly a larger version of Meteor Crater. The structure below the crater floor has been investigated by drilling which shows there is a filling of fall-back breccia (allochthonous breccia), below which the country rock has been strongly brecciated (autochthonous breccia). The allochthonous breccias form a crater-fill of moderately to weakly shocked material beneath which is a pool of melted rock, a remnant of material formed in the jetting phase as described earlier.

Brent is probably about the largest crater that can form as a simple bowl-shaped crater on Earth. Craters larger than this (e.g. Flynn Creek, USA (Roddy, 1963) and Clearwater Lake, Canada (fig. 5.8)) tend to develop central peaks (uplifts) (fig. 5.7), which become more prominent as the crater size increases. For example in Gosses Bluff in Central Australia (Milton *et al.*, 1972) the main form of the crater has been eroded away, leaving only the central peak. The rocks are Palaeozoic sediments but the crater itself was probably formed during the early Cretaceous. The Bluff is 4 km in diameter and consists of upturned strata dipping outwards in the form of a ring with the oldest rocks in the centre. It is essentially a circular anticline of rocks lifted some 3000 metres above their normal position below the surrounding plain. However, the outcrops of individual rock units in the sequence are not continuous around the Bluff, but consist of discrete plates

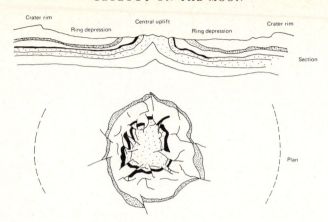

Fig. 5.7.   Cross-section and plan of rock strata in the central uplift of an idealized
terrestrial impact crater similar to the Sierra Madera of Texas (12 km diameter).

each a few hundred metres long. Close to the surface, the rocks are
more fragmented with individual unsorted clasts, randomly oriented.
In many places it appears that the vertical strata toppled outwards to
lie as overturned plates and detached blocks around the summit of the
uplift. Some of the blocks are so far from their original position that
they must have been propelled there by dominantly upward accelerations as uplift took place.

Outside the Bluff area, there are fragments of shock-metamorphosed
material consisting largely of sintered shock-melted fragments. This is
considered to be allochthonous breccia which was emplaced in the
floor of the original large crater.

The Ries basin impact structure in Germany (Denis, 1971) has a
diameter of 22–24 km (fig. 5.9) and was formed 14 to 15 million years
ago by impact into some 600 m of sediments underlain by crystalline
basement. The bulk of the ejecta is of sediments, although crystalline
rocks were also ejected and found within, or close to, the crater rim.
It appears that the ejecta has a reverse stratigraphy which, together
with shock metamorphic structures, confirms its impact origin.
Generally speaking, the size of the ejected blocks decreases away from
the crater. The bulk of the ejecta consists of the sedimentary material
and shows various stages of shock. The bedrock below the ejecta
breccias is often polished and grooved, with striations that are radial
to the crater. Near the crater rim are large slabs of limestone that may
be 25 to 1000 metres long. These are fractured and folded with axes
parallel to the crater rim (fig. 5.9), and it appears that slabs of the local

Fig. 5.8. Clearwater Lakes, Quebec, occupying impact basins formed by two meteoroids considered to have split apart from a single object on entry into the Earth's atmosphere. Each crater has a central uplift, the one in the larger West Lake being a ring. West Lake is 30 km across and was formed about 285 ± 30 million years ago. (Photograph: courtesy of M. Dence.)

rock were ejected from the crater and slid intact radially from the crater. This, together with the grooved and polished surfaces, suggests that once emplaced, ejecta still retains a radial velocity component causing movement along the ground away from the crater. The other type of ejecta at Ries is *suevite*, which consists of crystalline rocks in various stages of shock metamorphism and contains aerodynamically shaped bombs. The bombs appear to have consolidated before impact and most are about 2–12 cm across. An 84 m deep bore-hole through the suevite shows that the upper and lower regions are chilled, implying that suevite was formed in one episode during the impact event. Its fairly uniform composition suggests that most of it came from one well-defined source horizon in the bedrock. The degree of shock metamorphism corresponds to temperatures of up to 2000 °C when suevite was formed. Moreover there is evidence that suevite flowed after emplacement.

Ries is obviously important as an example of impact cratering and provides much information on impact processes and shock meta-morphism not seen so clearly in other terrestrial craters.

As we have seen, during a large impact, a small volume of rock may be melted. Although impact melt rocks could be (and have been) confused with igneous sills, dykes, and other volcanic rocks, their impact origin can be diagnosed by shock metamorphic effects. However in large structures relations are often not clear. For example, the Sudbury structure in Ontario appears to have been caused by an impact that unroofed a body of genuine igneous magma which was then emplaced within the crater. The structure is well known because large deposits of nickel, copper and iron are associated with the igneous activity.

Although we include Sudbury here as an example of an impact structure, this origin has not been confirmed to the satisfaction of all scientists who have studied it. Good evidence for impact is provided by shock metamorphism in breccia sheets on the basin floor (French, 1968b, 1970), and the presence of shatter cones (cone-shaped forms bounded by shear planes induced by shock). Igneous material occupies the inside of the basin between the breccia lens and the shocked bed-rocks and consists of norites, quartz diorites and micro-pegmatites making up a thickness of several kilometres. The impact took place about two billion years ago, possibly into sediments covered by a considerable thickness of water. French (1968) assumes that a central peak formed, but for such a large crater, many tens of kilometres across, a ring structure might have developed similar to those of large basins on the Moon. However, in French's central uplift model, the

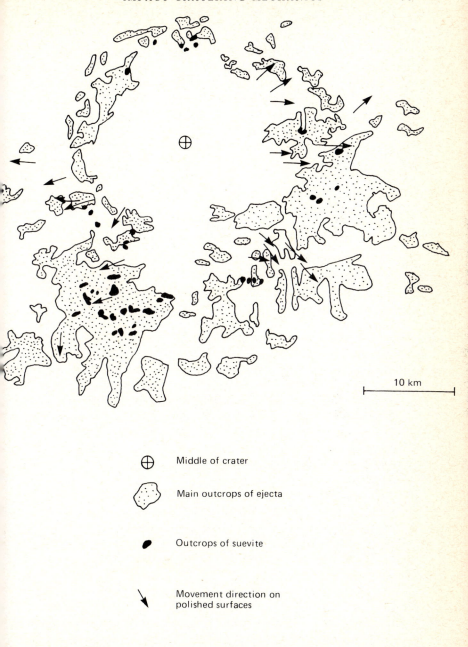

10 km

⊕ Middle of crater

    Main outcrops of ejecta

● Outcrops of suevite

↘ Movement direction on
    polished surfaces

Fig. 5.9. Distribution of ejecta, suevite and directions of striations on polished surfaces
at Ries impact crater, Germany. (Compiled from Denis, 1971.)

central peak subsided before the emplacement of the igneous rocks; magma was released along deep fractures below the crater and intruded below the floor of the crater along the contact between the basement rocks and the allochthonous breccia.

Despite some uncertainties about its origin, evidence for impact is strong and Sudbury is important because of its possible relevance to understanding lunar impact structures where volcanism has followed and been related to impact cratering.

5.4.   *Criteria for identifying impact craters*

The study of impact craters on Earth has led to a number of criteria for recognizing impact craters (Dence, 1972):

1. *The presence of meteoritic material near the structure.* This is rarely seen and there are some structures such as Flynn Creek (Roddy, 1968), where careful geochemical surveys have detected no evidence of meteoritic materials. Nevertheless, since most of the meteorite would be vaporized during the jetting phase, this is not surprising. In addition there are several types of object that could cause impact, including nickel–iron meteorites, chondrites and achondrites, as well as comets; as comets are considered to be formed of ices, they would leave no evidence of their existence.

2. *Circular plan of the crater.* Nearly all the moderately fresh impact craters, as well as nuclear and high explosive craters, are close to circular in plan view and show a much higher degree of circularity than almost all volcanic craters. Slight deviations from the circular in impact craters can result from bedrock control (see chapter 7).

3. *Uplifted and overturned strata in the rim.* This is always present in impact craters. With larger craters there may be a peripheral trough or ring syncline concentric with the rim.

4. *The central part* of the crater in smaller craters consists of a lens of *breccia* but with larger structures an *uplifted peak* of steeply dipping strata may be present or even a *ring of concentric* hills.

5. *Brecciation* is associated with the ejecta outside the crater as well as inside the crater, and the bedrocks surrounding the crater may also be locally brecciated *in situ*.

6. *Shock metamorphism* is one of the main criteria for recognizing impact craters, including dislocation of mineral lattices in varying degrees to the extreme form where the lattice is completely disrupted by high pressure; or alternatively melting from transient high temperatures has occurred to give glass.

7. *Large impact craters* are usually associated with negative gravity

anomalies and there may be distinct magnetic anomalies. Seismic velocities are usually lower than the surrounding country rocks.

Several points are particularly relevant to lunar impact craters. Impact melting clearly takes place and much of the melt remains in the crater. Greater quantities of melt compared to total ejecta will be generated by larger craters (Dence, 1971). It is also clear that at least for Ries-sized craters, ejecta, once emplaced, continues to 'flow' along the ground from the crater. These and other observations previously mentioned help in an understanding of lunar craters; but although the impact processes are the same on Earth or Moon, the six-fold difference in gravity will result in morphological differences between craters formed by similar impacts on the two planetary bodies.

6. Large lunar craters

6.1. *General morphology*

Fresh lunar craters show a range of morphology depending on size. In this chapter we are concerned with those craters greater than about 20 to 30 km in diameter and which show well-developed terraced inner crater walls, central peaks and ejecta patterns (figs. 6.1, 6.2). Although no crater of this class has been visited by a mission that returned samples, Tycho, which is probably one of the youngest large craters on the Moon, was visited by Surveyor VII, the last of the U.S. unmanned soft-landers. The relative youth of Tycho is indicated by the extensive bright ray system, the rays being seen to overlie most other geological units on the Moon.

Tycho's morphology is representative of most other large young craters, such as Aristarchus, Copernicus, Theophilus and Aristoteles. Its diameter is about 85 km and the crater floor lies some $4\frac{1}{2}$ km below the level of the rim crest. The central peak is well defined and rises about 2 km above the crater floor. There are also smaller hills nestled inside the crater. The inner walls of the crater are characterized by a series of terraces. Just outside the edge of the crater there is a narrow area of hummocky topography with relatively high albedo; immediately outside this area is a well-marked zone forming a dark halo around the crater and characterized by the radial lineaments (linear features) superimposed on broad undulations. This grades out into the bright ray system where numerous elongate craters form crater chains, crater clusters and loops of craters. Most other large lunar craters have all these morphological elements, although those that are somewhat older than Tycho may have lost some of their albedo markings such as bright rays by shallow erosion.

6.2. *Origin*

Although it is now generally accepted from the evidence of pictures from spacecraft that most large lunar craters are of impact origin, a number of workers have argued for an origin by volcanic–tectonic processes. McCall (1965) has made comparisons between lunar craters and certain calderas in East Africa. Green (1971) has also argued that they are calderas and has likened the rays to ignimbrites (ash-flows), while Fielder (1965) has explained them as complex volcanic extrusions building up rims along ring fractures. These arguments are based largely on morphological comparisons between terrestrial volcanoes and lunar craters, and as such do not stand up to rigorous morpho-metric analysis. In most cases there is an order of magnitude difference in the size of the terrestrial volcanic craters and lunar features being compared. Even making the unreasonable assumption that this could be explicable in terms of the different physical conditions on Earth and Moon, there are still other basic differences. For example, lunar craters, like terrestrial impact craters and experimental explosion craters, tend to be circular, whereas calderas do not. The raised rims of terrestrial volcanoes tend to be wide compared with the diameter of the crater, whereas lunar craters have relatively narrow raised rims, comparable to those of shock-wave craters on Earth (Guest and Murray, 1969). Also they are distinguished by the depth/diameter ratios and other parameters (Pike, 1974a, b). On the other hand, the geological characteristics of craters can be readily explained in terms of shock-wave cratering.

191622

Four individual large lunar craters, Copernicus (Shoemaker, 1960, Howard, 1975), Tycho (Shoemaker *et al.*, 1968), Tsiolkovsky (Guest and Murray, 1969) and Aristarchus (Guest, 1973) have been studied in detail. Shoemaker's arguments for an impact origin of Copernicus emphasize comparisons with nuclear craters. He was particularly impressed by the array of craters surrounding Copernicus in the bright halo areas (fig. 6.3) and these he likened to the secondary impact craters surrounding experimental craters.

Surrounding each of the big craters between the crater and the array of secondary craters (i.e. corresponding to the dark halo mentioned earlier for Tycho), is a rock unit that thins away from the crater and mantles the surrounding terrain. This unit is interpreted as fragmental ejecta that became draped over the surrounding terrain (rather than being emplaced as liquid flows such as lavas). The presence of rays extending for several thousand kilometres supports the concept of one gigantic crater-excavating event. Furthermore, lunar craters differ from most volcanic ones in having floors that are lower than the surrounding terrain. This and other evidence requires an origin as a single massive event capable of producing a crater some tens or even hundreds of kilometres across surrounded by ejecta thrown for many hundreds of kilometres; an extremely energetic event is required to throw large missiles (hundreds of metres across) for distances of some

Fig. 6.1. Vertical view of the 28 km diameter crater Euler, in Mare Imbrium showing
the nomenclature of a typical lunar impact crater. This crater is at the lower
limit of 'large' craters and has poorly developed inner terracing (wall slumping).
Compare with fig. 6.2. (NASA Apollo 17–2923.)

tens of kilometres to produce the secondary craters. Volcanic explosions
of the magnitude required to produce these effects could not take
place because the confining pressures within the lunar crust would not
allow sufficient pressure to build up to cause such an explosion (Roddy,
1968). The only mechanism that could produce such an event is a
shock-wave cratering event, and the only known natural process within

Fig. 6.2. Vertical view of part of the 100 km diameter crater Theophilus showing ejecta deposits. Continuous ejecta deposits (E) extend slightly more than one crater diameter outward from the rim and are characterized by terrain that is completely covered by material excavated from the primary crater. Continuous ejecta deposits give way to the discontinuous ejecta facies (S) where only parts of the terrain are covered and disturbed. The raised rim (R), terraced inner walls (T) and central peak (P) are also seen. (NASA Apollo 16–0432.)

the solar system that can induce such an event is a large impact. Further evidence in favour of impact is provided by the central peaks of lunar craters. These are typical of large impact craters and explosion

craters detonated near ground-surface; explosion events generated
internally (by some unknown volcanic process) are not likely to be
surface events.

6.3.   *Ejecta outside the craters*

Crater units outside the crater are of three types:
  (1) the continuous ejecta,
  (2) the discontinuous ejecta and bright ray material (fig. 6.1),
  (3) impact melt units (discussed in section 6.5).

Large craters are surrounded by a raised rim (fig. 6.2), the outer
boundary of which is marked by a distinct break in slope onto the
almost level portions of continuous ejecta. The raised rim is frequently
characterized by concentric fractures, although other surface characteristics may include hummocky terrain. On Tycho the hummocky rim
makes up only a small part of the whole circumference of the crater,
but on Aristarchus it represents almost half the rim of the crater.
Following our knowledge of shock-wave cratering, this raised-rim
unit was formed by bulk uplift and overturning of strata at the end of
crater excavation when the shock-wave velocity was too low to do
more than just lift rock in bulk out of the crater. It will therefore,
consist of local bedrock that has been moved only a short distance
from its original pre-impact position. Differences in rim morphology
may represent different lithologies in the excavated bedrock. For
example, the 40 km-diameter crater Aristarchus lies astride a fault
(fig. 6.4) that separates mare material from the older Aristarchus
Plateau material consisting of debris from the Imbrium event and other
units. Guest (1973) has pointed out that the hummocky rim unit occurs
on the Aristarchus Plateau side of the boundary fault but not on the
other side where there are mare materials and where the rim shows
normal concentric fracturing. On this basis, he has argued that a
difference in lithology between mare materials and rocks of the
Aristarchus plateau caused the different rim morphologies.

To continue with Aristarchus as an example, further out from the
crater's edge, continuous ejecta is characterized by radial lineaments;
and the thickness of the unit decreases away from the crater, as evidenced
by the way that features in the underlying topography, such as craters,
are progressively buried until they are no longer seen as the ejecta
blanket thickens close to the crater. For a crater the size of Aristarchus,
the maximum ejecta thickness is around 1 km thinning to some 30 m
where it grades into the discontinuous ejecta.

Concentric scarps on the surface of the continuous ejecta suggest
flow fronts and imply that the ejecta continued sliding over the surface

Fig. 6.3. The 100 km diameter crater Copernicus (bottom right) with well defined secondary craters and crater chains (arrows) outside the continuous ejecta. North is at bottom. (NASA Lunar Orbiter IV 121 H$_2$.)

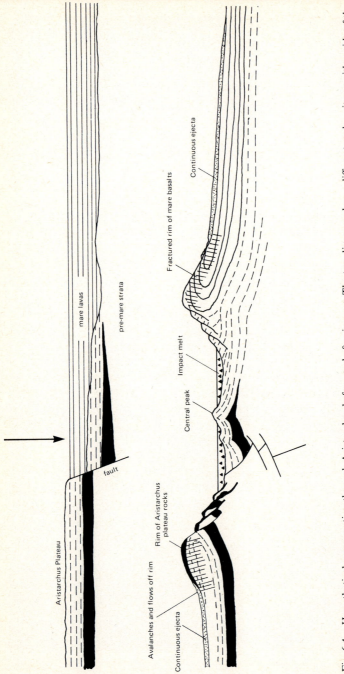

Fig. 6.4.   Hypothetical cross-section through Aristarchus before and after impact. The top diagram shows different rock units on either side of the Aristarchus Plateau fault. The lower diagram shows the possible deformation of strata below the Aristarchus crater which formed straddling the Aristarchus fault. Differences in lithology across the fault are thought to explain differences in rim morphology on either side of the crater. (After Guest, 1973.)

Fig. 6.5. Diagrammatic cross-section through overlapping sheets of crater ejecta (deceleration lobes) that have travelled forward after emplacement. Arrow shows direction of movement over old landscape. (From Howard, 1972.)

after emplacement (fig. 6.5) as a dense particle flow somewhat like an avalanche. At Aristarchus, Guest (1973) has suggested that continuous ejecta has been stripped off the rim unit leaving remnant patches. There are also radial grooves on the surface of the rim which appear to be scour-marks produced when the continuous ejecta was stripped off by some form of outward radial flow. Much of the ejecta, although thrown out before the overturning of the raised rim, was emplaced after completion of crater excavation and then flowed forward.

At the outer margin continuous ejecta grades into the bright ray system, and the boundary between these two units is difficult to determine (fig. 6.3). In this region the continuous ejecta thins and breaks up into discontinuous patches and appears to overlie secondary craters associated with the discontinuous ejecta; this is evidenced by the way that the secondary craters in the discontinuous unit become less distinct on the boundary with the continuous ejecta. For mapping it is convenient to take the subdued character of the secondary craters as a criterion for mapping the outer boundary of the continuous ejecta.

Secondary craters are often distinguished from small primary impact craters by their morphology, described by Shoemaker (1965). Primary craters tend to be circular and have a distinct raised rim, generally of uniform height. During the Ranger Missions, Gerard Kuiper commented that small primary craters are so uniform that they look as if they have been turned out on a lathe. Secondary craters, on the other hand, are less regular in form: they are often elongate and typically shallower than primary craters of the corresponding size. They have low irregular rims and often form tight clusters. However, while primary craters never look like this, some secondaries might resemble primary ones. Maximum numbers of secondaries occur at a radius of about one crater diameter from the primary crater. These swarms of secondary craters also often form characteristic crater chains and where these are radial to the primary crater, it is often seen that the crater at the end of the chain furthest from the primary crater appear to be the youngest crater on the chain, based on cross-cutting relations between craters.

The bright rays associated with these clusters of secondary craters probably consist mostly of locally excavated material with the difference in albedo representing a difference in texture from surrounding undisturbed terrain. Certainly the higher albedo material must be relatively thin since it is found only on relatively fresh craters. Slightly older craters show the same morphological characteristics but the rays have disappeared. Much of the ray material was probably formed during the jetting phase when material was thrown out at high velocities capable of carrying ejecta great distances, often as much as hundreds of kilometres from the primary crater, and on arrival churned up the surface to give rays. We have noted earlier that continuous ejecta appears to overlie secondary craters. Thus the secondary craters formed before the arrival of continuous ejecta by sliding and avalanching across the surface.

Another characteristic of the discontinuous ejecta region is the presence of herringbone ridge patterns and V-shaped ridges around secondary craters (fig. 6.6). These are particularly striking around Copernicus (Guest and Murray, 1971, Oberbeck and Morrison, 1973). Guest and Murray noted that there is a change in V-angle with distance from the primary crater, the angle increasing rapidly until about one crater diameter away from the rim of the principal crater and then remaining fairly constant at just over 100°. From experimental work at NASA Ames Research Center, Oberbeck and Morrison showed that when two simultaneous impacts occur close together, a ridge develops between the two craters as a result of collision of ejecta from each. If, however, the two simultaneous impacts are produced by impacting bodies with low-angle trajectories, then the resultant ridge between the two craters becomes V-shaped with the apex of the V pointing up-range. However, V-shaped ridges are also found on single craters, and in order to explain this, Oberbeck argues that the two simultaneous impacts were so close together as to form one single crater with a V around it. Laboratory experiments showing this phenomenon are less like the lunar case than the herringbone pattern. While the Oberbeck explanation goes a long way towards explaining these peculiar features, it may also be that there was some interaction between the forming secondary craters and material of the continuous ejecta moving outwards at high velocities. In this case, the V could be treated as a bow-wave phenomenon developed between the secondary crater and the over-riding ejecta from the primary crater (Guest and Murray, 1971).

Examination of any part of the lunar surface reveals numerous secondary craters originating from many different sources. Thus in any

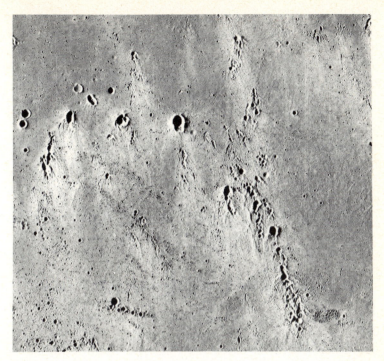

Fig. 6.6. Secondary crater chains and clusters from Copernicus, showing typical herring bone pattern made up of individual 'V'-features. (NASA Apollo 17–2289.)

region studied, the effect of secondary cratering must be taken into account. However, in terms of collecting lunar samples, it is unlikely that secondary cratering will produce strong contamination of the regolith by materials from the primary craters. The ejected blocks that produced the secondary craters must have been moving at very high velocity in order to have travelled the great distances involved; the volume of bedrock that they excavate will far exceed their own volume, so that contamination from any individual secondary block is small.

6.4. *Rock units and structures inside the crater*

The terraced inner walls of craters are thought to result from fault slumping, which has the effect of increasing the size of the crater. Although some slumping may have occurred long after the crater formed, most of it probably took place immediately after excavation as part of the impact cratering process. After excavation there is an attempt to regain the original surface morphology; the crater walls slump and the floor moves up and forms a central peak (fig. 6.4).

There may be a single peak (fig. 6.7) or a cluster of peaks standing one or two kilometres above the crater floor. Studies of high-resolution photographs of peaks such as those of Copernicus and Aristarchus show that they may consist of stratified rocks which have been folded and upturned in a manner to be expected if the peaks were formed in the same way as central peaks in terrestrial impact craters.

The floors of craters not modified by later mare fillings are covered with a layer of rock (fig. 6.7) having rugged, corrugated upper surfaces (cf. Orientale, Chapter 3). This unit overlies the terraces and is draped over the central peaks. There are three characteristics of the floor material that should be considered in relation to its mode of formation (Guest, 1973): (a) it has a ridged appearance resembling the patterns developed on viscous flows; (b) there is large-scale polygonal cracking of the surface, suggestive of contraction possibly caused by cooling; and (c) there are peripheral fractures on an inward sloping bench on the edge of the crater floor suggesting that the top of this unit was initially at a higher level and later subsided. From these observations, it may be concluded that floor material was emplaced as a hot, viscous layer which under lithostatic load contracted in its lower parts causing the surface to subside. Although it has been argued (Strom and Fielder, 1970) that the floor represents a large lava flow, it appears more likely that it is a mixture of impact melt and cold ejecta. Such material would be capable of sintering to form a compact, glassy rock that could flow and thus cause ridging, would become compacted (in much the same way as a thick ignimbrite or ash-flow) and could develop contraction cracks during cooling. Where it is thinner, it would be seen draped over hills on the floor of the crater.

6.5. *Flows and impact melts*

High resolution pictures of the rims of large, fresh craters show that in certain areas on the rim and inside the crater there are flow-like formations. These are particularly prominent on Aristarchus, Copernicus and Tycho, and King crater on the farside. There are three types of flow:

(a) longitudinally ridged flows;
(b) flows with ridges and levee-bounded channels; and
(c) pools and flows with dark smooth surfaces.

All of these flows have been interpreted by some authors as being volcanic (e.g. Strom and Fielder, 1971), but their close association with the rims of impact craters and the nature of the flows suggest that they may be associated with the impact event itself.

Fig. 6.7. Part of the floor of Aristarchus. The foot of the inner terraced rim (T) is in the lower part of the picture. Hummocky floor material (F), with cracks (C) overlies this and central peak material (CP). Other hills (H) project up through the floor material; these were probably thrust up by collapse of the terraces on the walls before the floor material was emplaced into its present position. (NASA Lunar Orbiter V 199 H_2.)

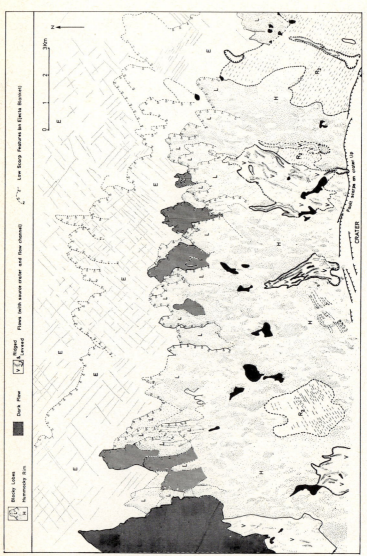

Fig. 6.8. Map of north rim of Aristarchus to show flows of various types. Note that blocky lobes and longitudinally ridged flows (*vertical ruling*) interdigitate. (From Guest, 1973.)

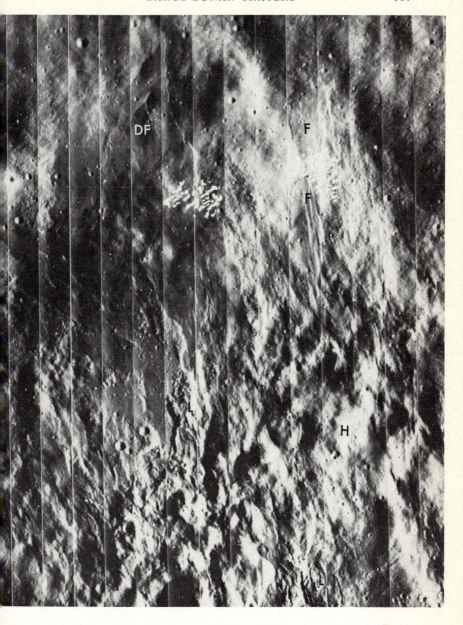

Fig. 6.9. High resolution picture of north rim of Aristarchus (far left of map in fig. 6.8) to show dark flow of impact melt (DF), longitudinally ridged flows (F), leveed flows (L) and hummocky rim material (H). (NASA Lunar Orbiter V 201 H$_3$.)

On Aristarchus flows occur on both the outer and inner parts of the rim and lie stratigraphically above the continuous ejecta sheet. The longitudinally ridged flows (fig. 6.8, 6.9) have lobate fronts and in many ways are like some pyroclastic flow deposits and certain types of landslides such as the Sherman Glacier landslide (Shreve, 1966), where the longitudinal features were formed by vertical shear planes developed parallel to the direction of movement in the flows. On Aristarchus, as the flows are followed up into their source area, they merge into the hummocky rim material with no well-determined source. The flows are interbedded with blocky lobes of material that appear to have slumped from the hummocky rim (Guest, 1973). The blocky lobes consist of poorly graded material and probably resulted from debris slides off the rim unit as it was emplaced. The observation that the smooth flows and the blocky lobes are interstratified strongly supports the view that the flows are formed during the Aristarchus event and are debris flows formed by avalanching.

The 200 km crater Tsiolkovsky (fig. 6.10) on the farside has one enormous 'flow' with well developed ridges and grooves parallel to the direction of the flow. The flow originates at the rim crest where it is characterized by a series of arcuate fractures concave down-slope. Photographs show that this part of the crater rim appears to have been pushed outwards along two short fractures, causing the outer wall of the rim to collapse and form this major landslide.

The other two types of flow associated with large craters appear to have been formed by liquids rather than fragmental flows (fig. 6.11). They range from rugged ridged flows with well developed flow channels bounded by levees, to smooth, thin flows emplaced in a very fluid state. The ridged, channelled and leveed flows appear in some cases to originate from craters, but this is not always the case. They usually occur high on the crater rim on the outside, and flow down the terraced walls on the inside. They may be three or more kilometres long and show quite complex patterns.

The smooth flows, on the other hand, are usually darker and may form extensive units that fill small hollows, or form narrow flows with small channels and lobate edges. They are often draped over the inner terraced walls and form pools at the back of the terraces. The broader sheets veneer rough topography and have complex contraction fissuring. It is quite clear from the evidence that this material originated as molten rock (Howard and Wilshire, 1975).

Characteristics such as the festoons of ridges and tension cracks indicate that the material flowed sluggishly like a lava flow or a wet debris flow, and not like avalanches or ash-flows. We know from lunar

Fig. 6.10.   Apollo picture of part of 200 km diameter crater Tsiolkovsky to show the avalanche on the rim (A). Tsiolkovsky has a rugged floor unit (R) overlain by dark mare material (M). The crater near top of picture has a flow-like unit on its floor (F); this may be volcanic, or more likely impact melt. (NASA Apollo 15–0892.)

samples that rocks on the surface of the Moon are anhydrous and it is unlikely that the flows were lubricated by water. It is much more probable that they were rock-melt that later solidified. There are several arguments that are against their being volcanic and are indicative of an impact melt origin. The first is that the thermal regime of

Fig. 6.11.   Downhill flow features and pools of lava-like material, interpreted to be impact melt, on north rim of King crater. Visible are flow lobes, flow channel, and fracture flow surfaces. Width of photograph is about 12 km. (NASA Apollo 16 panoramic frame 5000.)

the Moon changed with time and little volcanic activity appears to have occurred after about 2500 million years ago and yet we see these flows on impact craters that are thought to range in age over at least 3500 million years and continued long after volcanic activity elsewhere on the Moon had ceased. Also the various types of fluid material grade into one another and many appear to be draped on the topography rather than having flowed.

If these units are impact melts, then it would appear that rock-melt coated the crater floor and parts of the ejecta late in the cratering event and that flow continued after the slumping of the inner walls of the crater and after the radial flow in the continuous ejecta had ceased. Howard and Wilshire (1975) suggest that melt was splashed out as a mass, as evidenced at King Crater by some uphill facing flow-fronts, suggesting that these materials were propelled away from the crater. It is also likely that melt fell as a rain of hot bombs that coalesced and sintered together as they settled, providing viscous material that could flow.

6.6. *Summary*

All the features of large lunar impact craters can be explained in terms of shock-wave cratering produced by impact. On impact, high velocity filaments were thrown out in the form of an upturned cone of ejecta and the material from these filaments would have travelled for enormous distances across the lunar surface, producing numerous secondary impact craters. As excavation continued, the main bulk of the ejecta was thrown out until the end of excavation when the rim was formed by uplift and overturning before much of the earlier ejecta had fallen. Once emplaced, the continuous ejecta continued to move radially from the rim area in the form of fragmental flow. In these last stages, central peaks were formed by inward and upward movement of material below the craters and the inner walls slumped to form terraces. Hot breccia gave rugged floor materials, and impact melts of different viscosities continued to flow down the inner walls and over the continuous ejecta for some hours after the impact event.

The craters described in this chapter are all considered to be fresh and unmodified by later erosion and volcanism. Such modifications are considered in Chapter 9. Also, although the majority of large fresh lunar craters have this form, there are some with noticeable differences. One example is Kopff, near Orientale (fig. 3.10). This crater is younger than Orientale and is thus not old enough to have suffered major modification by erosion. Yet, although it has an ejecta blanket, it has no secondary craters. Were Kopff and craters like it formed by some unknown type of volcanism? Or were they formed by impact of a different type of projectile, or alternatively into a different type of bedrock, such as still hot and mobile Orientale ejecta? These questions remain unanswered.

7. Small craters

7.1. Introduction

Lunar craters smaller than about 15–20 km are much less complex than large ones (fig. 7.1). Their walls generally are talus slopes rather than having coherent slump blocks forming terraces on the inner crater walls as do their larger counterparts. Central peaks are usually absent. While small *impact* craters may be structurally less complex than large craters, it is in the small size classes that we find probable non-impact craters—craters formed by volcanic, tectonic or other endogenic processes.

Crater morphology has been of interest to selenologists for a long time, but it is only recently that accurate topographic data have been available for quantitative studies, particularly for small craters. Crater depth-to-diameter ratios for fresh craters (Pike, 1974 a) ranging in size from less than 100 m to more than 300 km show that craters smaller than 10–15 km are relatively deep compared with large craters. This can be explained partly by the differences in post-excavation modifications. As we have seen, the walls of large craters typically slump and the floors are often partly filled with impact melt—both processes that would make large craters shallower; these processes do not occur in small craters to the same extent. For small craters, the wall material is structurally sound enough to 'stand up', but when it does fail, it produces talus, composed of rocky debris.

7.2. Small impact craters

The outline of a crater in plan view gives a clue to its origin and the degree of modification. In general, crater outlines are described as circular, elongate, scalloped, or polygonal. Perfectly circular craters are almost completely restricted to those smaller than 15 km in diameter. Larger craters may also be circular, but close inspection of their rim crest shows that the outline is scalloped or polygonal. Scalloped outlines result from slumping of blocks of the rim, a process which not only causes an irregular crater outline, but may also increase

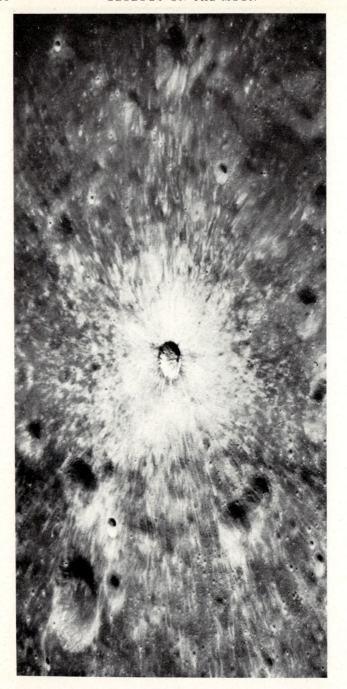

Fig. 7.1.   Typical fresh small crater (less than 10 km) showing bowl-shaped profile and bright ejecta patterns. Photographed by Apollo 17 astronauts from orbit with telephoto lens, west of crater Gagarin on farside. (NASA Apollo 17–150–23102.)

the diameter of the crater by as much as 20–40 per cent. As discussed above, slumping rarely occurs in small craters, and hence, the smaller craters generally do not display scalloping. Polygonal outlines for craters of all sizes appear to be the result of control by structures in the bedrock during the excavation stage or during later modification of the crater. Topography of the impacted surface may also affect crater shape (Schultz, 1976).

Several authors have used circularity as a means of interpreting the origin of craters, and many methods have been devised for assessing the circularity of craters (Murray and Guest, 1970). Circularity is useful for identifying primary impact craters, but a measure of circularity alone is not definitive. Because impact processes involve the rapid transfer of energy from the projectile to the target, an impact is a 'point source' event which excavates the crater in a uniformly radial direction from that point source—thereby producing circular craters. Departures from circularity are attributed to:

(1) structural control during excavation (producing, for example, polygonal craters),
(2) low angles of impact, producing elongate craters,
(3) post-impact modification (slumping of the walls, volcanic modifications, superimposed impact events, etc.),
(4) non-impact origin.

Elongate craters result from certain conditions of impact, or from endogenic processes. If the angle of impact for high velocity particles is less than about 5° above the surface, the projectile may plough through the surface, producing an elongate crater; on impact the projectile may even ricochet or break apart and produce a series of elongate craters. These craters normally have bilateral symmetry along the axis of the projectile's trajectory and show a distinctive ejecta pattern. Messier (fig. 7.2) displays these characteristics and compares well with laboratory simulations (fig. 7.3). In profile, elongate craters produced in the laboratory by low-angle impact events are deeper at the up-range end and the rim on the down-range end is higher. Secondary craters are often produced by missiles in low-angle trajectories and typically display some of these characteristics.

Strong departure from circularity and symmetry may indicate an endogenic (internal) origin, and more specifically, a volcanic origin. Analyses of terrestrial volcanic depressions such as calderas show that they are typically irregular both in profile and planimetric form. The reason for this irregularity is that most volcanic craters are the result of multiple events, in contrast to the single event of impact. Multiple volcanic eruptions seldom occur in exactly the same place, but instead

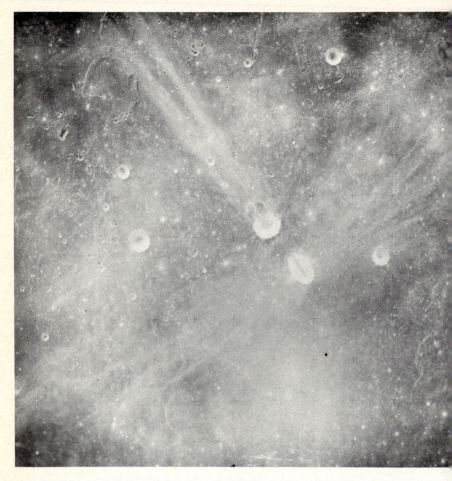

Fig. 7.2.   Vertical view of the twin crater Messier (~10 km diameter) and Messier A
produced by an oblique impact from lower right. This is a high sun angle photograph showing the asymmetry of bright ray patterns. (NASA Apollo 15–2674.)

shift their position. This results in craters with irregular outlines and
multiple-level floors. Volcanic pit craters, which result primarily from
collapse, are also generally non-circular, probably because of irregu-
larities in the competency of the rocks. If an impact crater were to
form in those same rocks, the irregular competencies would be completely masked by the high energies of the impact event and the rocks
would behave as essentially strengthless material.

There are many exceptions to the use of circularity to discern
impact craters from other types of craters. Any process that involves a

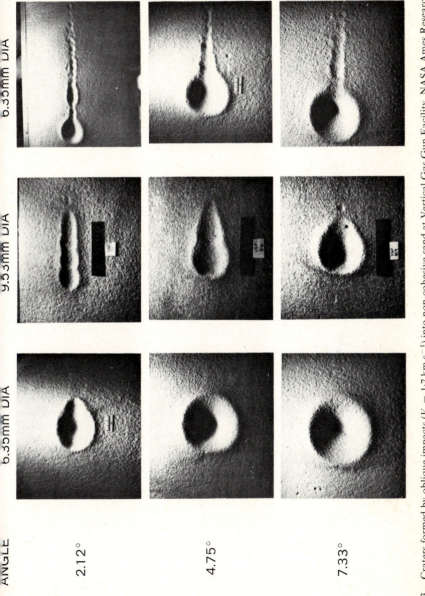

Fig. 7.3.   Craters formed by oblique impacts ($V_i = 1.7$ km s^{-1}) into non-cohesive sand at Vertical Gas Gun Facility, NASA Ames Research Center. (Courtesy D. Gault.)

sudden release of energy at a 'point source' could potentially produce a circular crater. For example, some volcanic craters known as maars result when rising magma intercepts water, producing a phreatic (steam blast) explosion. The resulting maar crater is circular, has raised rims, an ejecta field, and a floor whose level is below the surrounding terrain—all characteristics that are typically cited as indicative of impact events. Such a mechanism, however, requires the presence of water, which is unlikely to have been the case on the Moon. Nonetheless, the criterion of circularity, while useful in a general analysis of crater morphology, must not be used as the sole basis of origin identification.

The profiles of craters, that is, the topographic form of the crater floor, wall, and rim provide clues to the origin and degree of modification of craters.

Crater floors are described by their topography, surface texture, and surface features such as domes, pits, fractures, etc. Floors of small craters are particularly subject to modification by mass-wasting from the crater walls, by covering from adjacent exogenic and endogenic events, by volcanic processes and by tectonic events such as fracturing. Craters up to 5 km in diameter are bowl-shaped (with the floor and wall merging in a smooth curve), flat-floored, or have slightly domed floors. Bowl-shaped craters are thought to have resulted from impacts that excavated only regolith or other unconsolidated rock. Thus where the regolith is thicker, for example in the highlands, bowl-shaped craters have a larger maximum size.

Flat-floored craters are common in all size ranges. In craters smaller than about 1 km, this geometry appears to result from impact into a regolith-covered competent substrate, in which the excavation removes the regolith, but has relatively little effect on the substrate. This type of cratering is discussed more fully in the next chapter. The flat floors of some craters, such as those lacking well defined ejecta fields, but which have blocks on the floor, may have resulted from partial burial by an adjacent event such as an impact, or from volcanic eruption. The textures of flat floors provide clues to post-impact modifying processes. Hummocky floors can be the result of mass-wasting in the form of debris flows from the walls. Some flat floors have smooth, mare-like textures that may represent post-impact volcanism.

Domed, or convex, crater floors may represent an incipient central peak development, volcanic modification, or floor rejuvenation. For certain craters, volcanic processes appear to be the most likely, especially if the crater occurs in a suspected volcanic region. The convexity in these cases represents large, dome-like hummocks which either are randomly arranged on the floor, or occur in distinct patterns. In other cases, the convex floors could represent an upwarp, or rejuvenation of the crater floor through some sort of rebound or isostatic adjustment, or perhaps even intrusive activity.

The talus walls of small craters extend from the rim crest, or near the crest, to the floor. Typically, rock outcrops are exposed in craters 1 to 20 km in diameter. The profile of the wall may be straight, concave, or convex, depending, at least in part, on the mode of mass-wasting of the wall slopes, the physical characteristics of the fragments, and the thickness of the regolith in relation to the size of the crater. Walls of craters smaller than 1 km in diameter commonly exhibit blocks and a coarse rubble (fig. 7.4), especially if the crater was excavated in rock with a thin regolith covering, or if the crater was formed by a low-velocity impact in which the projectile was left intact or only slightly fragmented.

Small craters show the effects of degradation more rapidly than their larger counterparts. Degradation occurs through repetitive meteoroid bombardment, mass-wasting, and blanketing either by ejecta from adjacent impact events or by volcanic extrusions. The craters pass through progressively subdued stages until they are obliterated. Mass-wasting appears to be the prevailing process in the degradation of the walls of craters 1 to 20 km in diameter; surface creep appears to be responsible for producing a type of lunar surface texture called 'tree-bark' pattern (fig. 8.2); avalanching produces talus cones, boulder trains and trails, and blocky slides.

Rapid mass-wasting of the crater walls through avalanching is evident by the rejuvenation of wall features, such as outcrop zones, and the absence of small craters, while the ejecta field of the parent crater becomes subdued.

One way to create rapid mass-wasting is seismic triggering by impact events. Recently, Schultz and Gault (1975) demonstrated that the seismic energy released by large impact events was sufficient to cause substantial surface deformations, even at distances far from the source. Such a mechanism is undoubtedly very effective in initiating avalanches on slopes which are at or near the angle of repose.

Ejecta patterns of small fresh impact craters may include radial grooves, strings of blocks (fig. 7.6) and rays. With age, meteoroid bombardment and other processes degrade these features, making them progressively subdued. There are, however, many anomalies to this sequence: for example, there are some subdued craters that have sharp, well-defined blocks on their rim. This and other variations in appearance of small craters may be the result of one or more variables,

Fig. 7.4.   View across the floor of a small crater at the Apollo 17 landing site in the Taurus-Littrow valley, showing its blocky floor and fragmented regolith exposed in the far crater rim. (NASA Apollo 17–137–21002.)

including differences in sub-surface structure or rock competency, impact velocities and variations in degradational processes.

7.3.   *Dimple craters*

The first close-up pictures of the Moon were obtained by U.S. Ranger spacecraft. These pictures revealed numerous craters that have funnel-shaped profiles and which lack raised rims. First described by Kuiper (1965), these craters were named dimple craters. Although most are

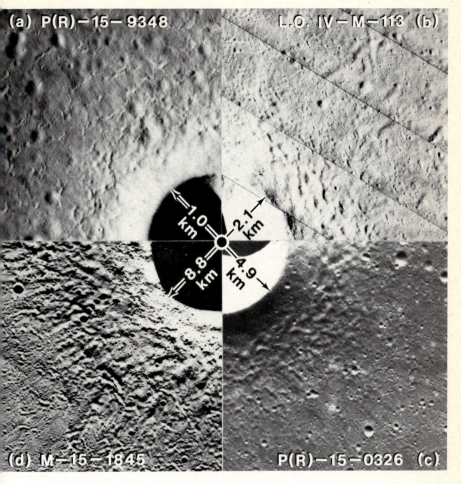

Fig. 7.5. Composite photograph showing changes in ejecta patterns for small fresh craters as a function of size. (Courtesy V. Oberbeck.)

smaller than about 500 m in diameter, some are as large as 1 km across; regardless of size, ledges of rock crop out in the walls of many dimple craters.

Several different modes of origin have been proposed for dimple craters: (1) they may be endogenic landforms that resulted from the collapse of plastic crust on partly cooled basalt flows, similar to dimple shaped depressions observed in many Holocene terrestrial lava fields (e.g. Greeley, 1969); (2) they may be secondary endogenic landforms that result from the drainage of surface regolith or other fragmental material into subsurface cavities such as fissures, fractures or

Fig. 7.6. Blocky-rimmed crater about 590 m in diameter in Oceanus Procellarum. Individual ejecta blocks, some more than 25 m across, are visible on the rims. (NASA Lunar Orbiter III 189H.)

lava tubes; (3) they may be degraded versions of concentric impact craters (Oberbeck, 1970). All three mechanisms appear to be workable hypotheses in view of our present knowledge of the lunar surface. Marial basalts were very fluid during their emplacement and, at least in some cases, seem to be analogous to the types of terrestrial lava flow in which collapse depressions form. In addition, the existence of lunar lava tubes and crustal fractures is generally accepted and non-indurated regolith would be prone to mass-wasting into such cavities. One constraint on this model would be that a given thickness of regolith would not be able to produce a crater above a certain size. Assuming the regolith to be entirely impact-generated this maximum diameter is about 44 m (Oberbeck, 1970) for craters in the maria, and this suggests that the third hypothesis might be more appropriate, at least for larger dimple craters.

As with so many landforms, it is probable that lunar dimple craters can result from any one of several processes and craters must be assessed individually on the basis of morphology, geological setting and apparent age.

7.4.   Crater chains

Several categories of crater chains (or *Catena*) are found on the Moon. The most common crater chains are strings of secondary craters, which can be identified by their orientation to larger primary craters, or by the presence of V-shaped ridges extending from the crater chains, described in Chapter 6. Other chains of craters, however, appear to be volcanic or tectonic in origin.

The Hyginus rille (fig. 7.7) consists of a series of craters several kilometres in diameter along a linear graben-like rille. The craters are rimless and, from their arrangement, it is unlikely that the chain of craters resulted from impact cratering.

Figure 7.8 shows a series of elongate craters, each several hundred metres across, that alternate with domical structures in northern Oceanus Procellarum. This crater chain has been described as resulting from partial collapse of a lava tube (Oberbeck *et al.*, 1969). Several other chains of elongate craters on the Moon have been attributed to this mechanism.

Not all chains of craters are so neatly categorized by origin, however. Davy Catena, for example, is often cited as an example of a series of lunar maar craters. In fact, during the Apollo series, this crater chain was being considered as a landing site because of its potential yield of deep-origin volcanic material. Recent examinations of high resolution photography for Apollo 16 and detailed topographic maps suggest

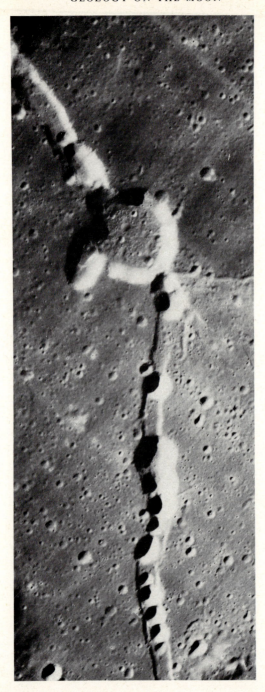

Fig. 7.7. Hyginus rille, a graben structure with collapse pits. The largest pit has a diameter of about 10 km. (NASA Lunar Orbiter V 96 M.)

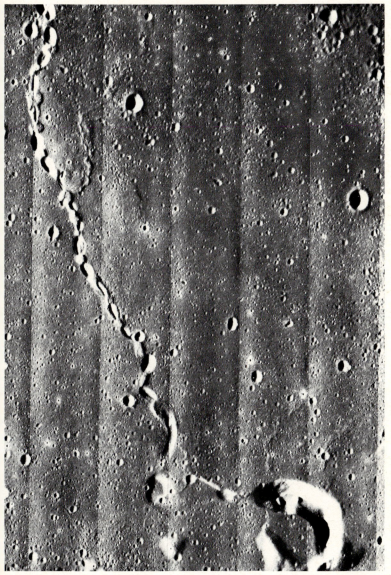

Fig. 7.8.   Crater chain (individual craters $\frac{1}{2}$ to 1 km in diameter) in northern Oceanus Procellarum, considered to be a partly collapsed lava tube. (NASA Lunar Orbiter V 182 M.)

that Davy Catena may be a chain of secondary craters, although its primary crater-source has not been identified.

7.5. Dark halo craters

Dark halo craters are characterized by a zone around the crater rim that has a lower albedo than the surrounding surface. These are relatively rare lunar craters that are typically less than 5 km in diameter.

On the floor of Alphonsus (fig. 7.9) dark halo craters, 2 to 3 km across, are often elliptical and sit astride fractures or graben which they appear to fill. It is likely that these and some other dark halo craters result from volcanic processes in which the 'halo' is pyroclastic ejecta. However, other dark halo craters that in all other respects resemble normal impact craters may owe their dark halo to dark material excavated from below a thin mantle of high albedo material.

7.6. Ring-moat structures

Several mare regions of the Moon are characterized by a high frequency of ring depressions (Schultz, 1976 and fig. 7.10). The depressions, or moats, may occur as simple rings a few tens of metres deep on the mare surface, or they may encircle small domes, some of which have summit depressions. The rings may be up to 1 km across, while the ring-moat domes are usually less than 500 m across. Comparisons with terrestrial analogues suggest that some lunar moats are annular depressions between a flow front and a topographically high obstacle on the pre-flow surface (Schultz and Greeley, 1976). Some may be sags in the lava surface caused by sub-surface lava drainage or reflect craters buried below the lava.

7.7. Crater counting—clues to the past

A recurring problem in geology is the need to date different rock units. Geological dating is either 'relative', in which units are older or younger than other units, but with no indication of how much older or younger, or 'absolute', in which the age of the unit is expressed in years since its formation. Relative dating is based on geometrical relations such as superposition and cross-cutting relations, while absolute dating is usually based on isotopic methods.

Several techniques have been derived for obtaining both relative and absolute ages of lunar surfaces based on crater 'counts', or crater-frequency distributions. Most of these techniques are based on the premise that, first, most of the craters on the lunar surface are impact and, second, that the longer a surface has been exposed, the more craters it will display. Figure 7.11 is a series of pictures taken of a sand

Fig. 7.9. Oblique view southwards over Alphonsus, a 117 km diameter, pre-Imbrium crater modified by Imbrium sculpture. The floor of the crater is cut by rilles on which are elongate volcanic craters surrounded by dark ejecta (arrows). (NASA Apollo 16–2478.)

box that was 'cratered' in the laboratory and illustrates the concept. With time, the surface becomes more heavily cratered and has larger craters. It is possible to count the craters on one frame and compare the crater frequency with another frame to determine which is the older surface. By extrapolation to the Moon, we can see that the lunar highlands are more heavily cratered than the marial plains, and hence are older, a conclusion supported by superposition of the maria.

Fig. 7.10.   'Ring-moat' feature several hundred metres across, considered to be a primary structure developed on the mare-lavas. (Schultz *et al.*, 1976.)

Taking the premise one step further, if one knows how often craters of a given size were formed on the surface, then it ought to be possible to count and measure the craters, divide the number formed per unit time, and obtain an absolute date for the surface.

There are, however, a number of problems with dating surfaces by crater counting. For example, in theory, there is a limit to the number of craters that can fit on any given surface. In other words, the surface is geometrically *saturated* (Gault, 1970) with craters. In reality, however, surfaces never attain geometric saturation because of the random placement of impacts. A more useful concept is that of *equilibrium* (fig. 7.12), in which craters are destroyed at the same rate that they are formed. This concept is illustrated by fig. 7.11. Close examination of the last row of frames shows that, while the arrangement of craters shifts position, the actual number of craters of each size remains essentially the same. This surface, then, is said to be in equilibrium and neither relative dating nor absolute dating would be possible.

Another constraint on crater counting techniques is the possible mixture of non-impact craters. Such a mixture would make a surface

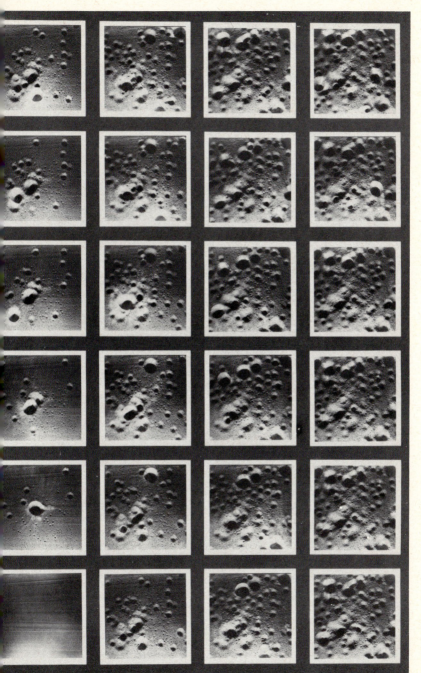

Fig. 7.11. A laboratory simulation of development of cratered terrain. This is produced by impacting six different sizes of projectiles such that ten craters were formed in one size class, for one crater in the next larger size. With time, the number of craters formed in one size class equals the number destroyed. (Courtesy D. Gault.)

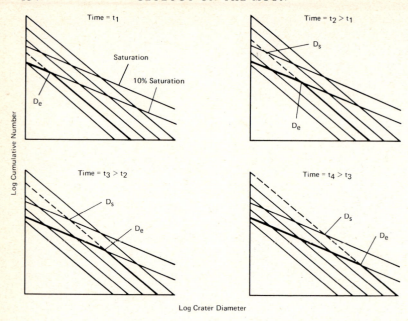

Fig. 7.12. Schematic representation of ideal changes in time of size-frequency crater
distribution on the lunar surface. The numbers of craters formed are given by
bold lines and thin dashed extensions. However, in practice the 10 per cent
saturation curve defines the maximum number of craters present of any given
size. When the crater frequency line meets the 10 per cent saturation line it thus
turns and follows it. (From Gault, 1970.)

appear older. There are many possible sources of non-impact craters,
such as volcanic collapse depressions and drainage features, particularly
for craters smaller than about a kilometre in diameter. 'Abnormalities'
in the crater counts may signal the presence of endogenic craters, as
seen in the 'pools' on the interior rim of Copernicus discussed by
Greeley and Gault (1971).

On the other hand, destructive processes would cause a surface to
appear younger. For example, seismic 'jostling' might degrade some
craters beyond recognition, so that when the crater frequency for the
surface was obtained the count would appear erroneously young. A
more complex case would occur when a surface might be blanketed
with a thin deposit that would bury small craters, but would only
partly degrade large craters. Crater counts would show essentially two
distributions and it would be difficult to separate one set of data from
the other.

Despite the problems with crater counting, within certain constraints it can be a useful aid in deciphering lunar stratigraphy and surface geology. In comparing one surface with another, large, homogeneous areas for each unit must be counted; the type of terrain should be similar; the images used to determine crater frequencies for both units should be of the same quality (resolution, angle of illumination, etc.); the same people should do the crater counting on both units; and craters large enough to eliminate possible problems with endogenic craters should be used. Within these constraints, it is possible to obtain relative ages for some lunar units.

Derivation of absolute dates for lunar surfaces by crater counting is more difficult. In the first place, one must know fairly accurately the flux of bodies that are potential impactors, and it must either be assumed that the flux has been constant over geological time, or the flux variations must known. At present, knowledge of meteoritic and cometary fluxes is poor. If one uses 'straight' crater counts (counts of all observable craters), then one must separate primary craters from secondary craters, a rather chancy business at best. Because of this problem, Soderblom *et al.* (1974) derived a technique of looking at the 'decay' of a crater profile. The walls of small craters become less steep with time, and through a statistical analysis of crater wall slopes, absolute dates could be derived, again with some knowledge of flux and cratering mechanics. However, the problem of non-impact craters again crops up, especially where small craters are involved.

The Apollo samples yielded isotopic ages for six lunar surfaces. Crater counts on these surfaces have been 'calibrated' to these dates (fig. 9.9), thus circumventing some of the uncertainties with the flux. The span of geological time represented by dated surfaces, however, is only from about 3200 to 3800 million years and extrapolations to both older and younger surfaces must be viewed cautiously.

8. Erosion, regolith and shock metamorphism

8.1. *Introduction*

There is no erosion by flowing streams, rain, or wind on the Moon at present, nor does there appear to have been in the geological past. Nevertheless, gradation does occur on the Moon. Gravity is an effective (albeit slow) means of moving material down slopes, even without the lubricating action of water. Mass-wasting is present in many forms on the Moon, ranging from surface creep to debris avalanches (fig. 8.1). The great range of lunar surface temperatures probably enhances thermal creep of surface material and may be the primary agent producing the tree-bark texture seen on many slopes (fig. 8.2).

Lunar seismic events also have an effect on gradational processes. Large-scale seismic effects accompany basin-forming impacts and smaller ones undoubtedly occur in association with smaller impact events, probably resulting in jostling of fragmental surface material downslope to degrade surface features.

The most important mechanism of degradation or erosion on the Moon is primary and secondary impact cratering. Impact craters range in size from more than 1300 km for the Imbrium basin down to micro-craters (fig. 8.3) only microns in diameter. In contrast to the Earth, where the smaller meteorites are consumed by the atmosphere, on the Moon objects of all sizes strike the surface and the countless billions of impacts have fragmented the lunar surface, causing a redistribution of surface material. The longer any given surface is exposed to impacts, the more degraded it becomes.

The net result of the rock-smashing impacts is to generate the mantle of rock fragments, known as the *regolith*, that blankets the lunar surface. Short (1975) enumerates the evidence for an impact origin of regolith:

1. Incorporation of fragments of underlying basalt.
2. Small mean particle sizes ($<60\,\mu$m) and correlation of size-distribution with extent of turnover.
3. Wide range of expected shock features.
4. Shock-lithified breccia fragments.
5. Glass spatter from shock melting.

Fig. 8.1. Vertical view of the Apollo 17 landing site (X) in the Taurus-Littrow valley between South Massive and North Massive, showing prominent mare-type ridge that appears to be a thrust fault (arrow), and a distinctive light mantle (M) that extends from the South Massive. This mantle appears to be a landslide deposit. Repeated landslides like this could contribute to the accumulation of regolith at the base of mountains. Width of picture about 25 km. (NASA Apollo 17–150–23005.)

6. Presence of nickel–iron meteoritic debris.
7. Microcraters on rock fragments and glass surfaces.

A process that goes hand-in-hand with impact cratering and regolith generation is *shock-metamorphism* in which rocks and minerals undergo

Fig. 8.2.   Hill creep giving 'tree-bark' texture and mass-wasting on hills of Flamsteed P in Oceanus Procellarum. Width of picture about 10 km. (NASA Lunar Orbiter I-199 M.)

chemical and physical changes as a result of the very high transient temperatures and pressures during impact. Under some circumstances, regolith can be lithified to produce breccias (fig. 8.4), and in fact, many lunar rock samples consist of shock-lithified rocks that are composed of

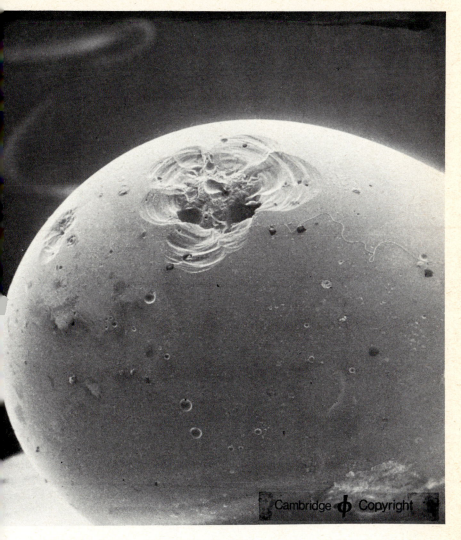

Fig. 8.3.   Microcrater on glass bead from regolith at Apollo 11 site. The crater is less than 0·3 mm across. (Courtesy S. O. Agrell, Copyright Cambridge Electroscan.)

breccias-within-breccias-within-breccias—indicating their multi-stage evolution.

8.2.   *Regolith*

The lunar regolith is a layer of debris consisting of virtually non-sorted particles ranging in size from microns to tens of metres across. The

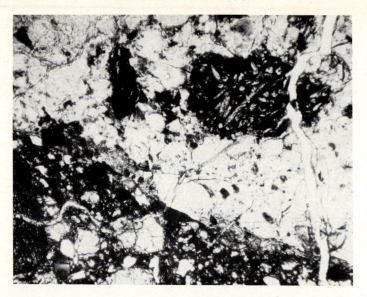

Fig. 8.4. A photomicrograph of regolith breccia collected from the Apollo 15 site. The dark area to top right is a clast set in a glassy mass containing glass spherules. To lower left is fragmental ground mass. The scale is 2 mm from left to right. (Photograph courtesy A. L. Albee.)

finer fraction is often termed lunar soil, and samples were collected during all the Apollo missions. They generally consist of the following components (from Warner, 1975):

1. Fragments of bedrock;
2. Fragments of rock and glass not like the bedrock and presumed to be derived from remote sources;
3. Agglutinates of particles bonded by glass droplets;
4. Glass droplets in a variety of forms and of both impact-melt and volcanic origin;
5. Meteoritic material.

The material making up the regolith is derived primarily from local sources, but also includes impact projectiles and mass-wasted debris, such as material released from slopes by seismic triggering. Returned samples show that only a few per cent of material is meteoritic. In some areas, especially those of high relief, material derived from surface creep and landslides may constitute a significant fraction of the regolith, as seen at the Apollo 17 landing site (fig. 8.1) where a tongue of debris spreads across the mare plain for many square kilometres.

Another possible source of regolith material is from volcanic pyroclastic activity. Although the amount of pyroclastic activity on the

Moon is not known, samples returned from several sites show particles of orange and emerald green glass that are probably of volcanic origin because of their uniformity of composition, high volatile content and similarity in age to the mare materials. Moreover photogeological studies of other sites, such as the Marius Hills, indicate pyroclastic landforms. Thus, in at least some areas, volcanic particles are mixed with the regolith.

By far the largest fraction of material in the regolith is derived from local bedrock by impact processes, demonstrated by Oberbeck *et al.* (1973) and Gault *et al.* (1974) through crater experiments and analyses of ballistic ejecta trajectories. Only about 5 per cent of the regolith would come from distances greater than 100 km, while more than 50 per cent is derived from less than 3 km (Taylor, 1975); these figures, however, are dependent on the age of the unit being impacted.

The physical properties of the regolith were of great concern in planning lunar landings. Some investigators were afraid that spacecraft might sink in a thick layer of fine-grained fluffy dust. Consequently, a great deal of attention was devoted to the engineering properties (bearing strengths, etc.) of the lunar soil that could be determined by unmanned Surveyor spacecraft (fig. 8.5). From Surveyor and subsequent Apollo missions, the dust problem was laid to rest—the lunar soil and regolith was found to be a rather stable mass of fragments with only the upper few centimetres being subject to compaction of the slightly cohesive fine particles (fig. 8.6).

The relatively high compaction in the regolith caused some problem in attempts to obtain even shallow core samples during the Apollo landings. Early attempts yielded cores that were too disturbed to be interpreted. It was not until Apollo 15 that a rotary percussion drill was used successfully to obtain core tubes of essentially undisturbed samples of the upper regolith. The longest core penetrated 242 cm into the regolith and revealed forty-two discrete layers, most of which probably represent overlapping ejecta deposits from successive impacts. The density of the core tube increased from 1360 kg m^{-3} in the upper half to 2150 kg m^{-3} in the bottom half, indicating the greater compaction with depth in the regolith.

Early in the Apollo programme, the term lunar 'gardening' was applied to the concept that the regolith was undergoing churning or overturning and many investigators gained the impression that the gardening occurred on a fairly rapid time-scale. Recent calculations by Gault *et al.* (1974) and others, however, show that most of the turnover of the lunar regolith takes place in only the upper 1 mm. At 99 per cent probability they show that the upper 0·5 mm is turned over almost 100

Fig. 8.5.   U.S. unmanned soft-landing craft, Surveyor III. In the background is the
Apollo 12 Lunar Landing Module. (NASA picture.)

times in a million years, but it would take 10 million years for the
upper 1 cm to be turned over just once. They calculate that the material
in the bottom of the 242 cm Apollo 15 core tube must have lain
undisturbed for 500 million years!

   The turnover of the upper millimetre or so of regolith and the
production of lunar soil (the fine-grained component) is primarily the
result of micrometeoroid bombardment—arbitrarily defined as those
impact events that create micro-craters roughly 1–20 mm in diameter

Fig. 8.6.   Surveyor III footpad area photographed by Apollo 12 astronauts, showing footpad prints formed during a touchdown hop. Fine-grained cohesiveness of the regolith is evident by the pattern of the footpad honeycomb and the steep walls of the depression. (NASA Apollo 12–48–7110.)

(fig. 8.3). Micro-cratering produces two effects on the lunar surface, discussed by Hörz *et al.* (1975). The first is a sand-blasting effect in which tiny chips of rock are removed from exposed surfaces as a result of impact events that are small in relation to the target. The second is 'catastrophic rupture' of exposed rock occurring when the impact is large in relation to the target. Using estimates for the parameters

Fig. 8.7.   The regolith surface at the Apollo 17 landing site in the Taurus-Littrow region, showing boulders with rounded surfaces and debris at base of boulders caused by erosion of boulder by micro-meteoritic bombardment. Lunar Rover in background gives scale. (NASA Apollo 17–138–21048.)

involved, Hörz *et al.* (1975) calculated an erosion rate for crystalline lunar rocks of 0·3 to 0·6 mm per million years; rates for breccias probably would be higher because of their lower compressive strength (Gault *et al.*, 1972). Thus, when we see a picture of the lunar surface such as fig. 8.7 we know that it has taken many million years to produce the crumbled debris that surrounds individual boulders, in contrast to Earth, where such a surface could be produced on a much shorter time-scale.

Regolith is also produced by bigger impacts. As discussed in chapters 3 and 6, large impact events shatter the lunar crust, produce continuous ejecta deposits over large areas and release significant seismic energy. Such events are very effective in producing regolith but occur rarely.

Thus, micro-cratering affects only the upper few millimetres and 'megacratering' has been too rare since the period of heavy bombardment to produce lunar-wide regolith. It is, then, left to cratering in the size range ~10 to 1500 m to generate regolith. Quaide and Oberbeck (1975) have used a highly refined regolith growth-mixing model (substantiated by photogeological studies of the lunar surface) to show that the accumulation of regolith is initially rapid, followed by diminishing rates of growth. This is because the regolith becomes an effective barrier against subsequent growth by impact excavation as it thickens. Thus, new material is added to the regolith only by larger impacts, which occur less frequently. With time, larger and larger impacts are required to reach through the buffering regolith to excavate bedrock, and the rate of regolith-growth falls. Although mixing increases with time, it never reaches the point where the regolith is homogenized, as the layered core tube samples demonstrate.

Several methods for determining the thickness of regolith in mare regions have been devised. The best method is to observe the thickness of the fragmental layer where it is in contact with the bedrock and exposed to view, such as in crater walls or along the rims of rilles. Such observations are valid only for limited areas and may be confused by slumping and mass-wasting, which often obscure the contact.

Data from the Apollo seismic stations and radar sounder instrument flown in orbit on the command module of Apollo 17 have yielded information on regolith thickness, but interpretations of the seismic data and the sounder for comparatively shallow depths are difficult, probably because the contact between regolith and bedrock is not a sharp, well-defined plane: rather the contact may be gradational in nature, with the fragmental regolith giving way to blocks of bedrock *in situ*, underlain by less fractured bedrock. The problem is made even more difficult in mare regions where voids and fractures in basalts might be expected, and in the highlands where possibly a thin mare-type regolith overlies a more massive, compacted 'mega-regolith' consisting of breccias associated with basins and large craters (Hartmann, 1973).

A method based on the morphology of small craters has been relatively well tested for determining mare regolith thickness (Quaide and Oberbeck, 1968). In essence, a crater excavated entirely in deep regolith is bowl-shaped, while an impact of equivalent size in thin regolith covering 'bedrock' will produce a crater with an inner terrace

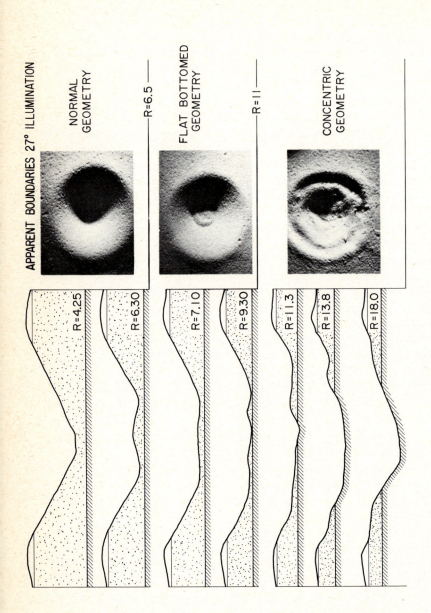

APPARENT BOUNDARIES 27° ILLUMINATION

NORMAL GEOMETRY

R=6.5

FLAT BOTTOMED GEOMETRY

R=11

CONCENTRIC GEOMETRY

R=4.25
R=6.30
R=7.10
R=9.30
R=11.3
R=13.8
R=18.0

Fig. 8.8. Diagrams illustrating the relation of crater geometry to the thickness of regolith overlying a consolidated substrate. (Courtesy V. R. Oberbeck and W. L. Quaide.)

known as a concentric crater because of the unequal excavation of the loose material giving a large outer crater and more competent underlying material giving a small, inner crater (fig. 8.8). An intermediate regolith thickness would result in a flat-floored crater. Since the initial work, the authors have substantiated their method with statistical analyses and photogeological studies of various mare surfaces.

By measuring and counting the different types of craters, it is possible to determine the range of thicknesses and the average thickness of the regolith in any given area. Although the thickness is variable in any one area, there is a good correlation of average regolith thickness with mare surface age, older surfaces being thicker. These relations were predicted before the Apollo landings for several sites and essentially confirmed with results from the missions.

8.3.   Impact metamorphism

Even before impact cratering was accepted as a geological process, unusual petrographic effects were recognized in meteorites and in samples from some terrestrial craters. For example, in 1872 Tschermak, in his work on the Shergotty meteorite, described maskelynite, an isotropic form of feldspar now accepted as a product of shock metamorphism. Around the turn of the century, unusual forms of the Coconino sandstone from Meteor Crater, Arizona, were described in great detail; we know now that those forms also were the product of shock metamorphism.

In Chapter 5 we discussed impact cratering and described it as a shock-wave process in which there is a rapid transfer of kinetic energy from the incoming object to the target—the planetary surface. A shock wave passes through the rock, subjecting it to high pressures and temperatures—the two 'agents' of metamorphism. It is reasonable, therefore, to expect those rocks to be altered, or metamorphosed.

Shock metamorphism involves the changes that take place in rocks and minerals by the passage of transient high-pressure shock-waves, over a period of time of microseconds to less than an hour or so. The only known natural mechanism for this process is very high velocity impact (hence the term *impact metamorphism*), but the same effects can be produced in the laboratory and in nuclear and chemical explosions.

How, then, does impact metamorphism differ from normal metamorphism? French (1968a) outlined several differences:

1. Very long time intervals are involved in normal metamorphism, while fractions of a minute, or at most a couple of hours, are involved with impact metamorphism even for very large impact events;

2. Temperatures in normal metamorphism seldom exceed 1000 °C while temperatures from impact may locally reach 10 000 °C;

3. Pressures during normal metamorphism generally do not exceed 50 kilobar and are usually lower than 10 kilobar, while impact peak pressures may reach several megabar. Because the reactions are so rapid and the quench times so brief in impact metamorphism, we find that chemical and physical *disequilibrium effects* are dominant, in contrast to normal metamorphism.

Impact metamorphism is so unlike any other geological process that its effects are important signals for the recognition of impact processes. There are three general changes that can occur in rocks and minerals subjected to impact metamorphism (French, 1968a):

1. *High pressure effects* which result in mineral polymorphs such as coesite and stishovite (high pressure forms of quartz) and diamond. Fig. 8.9 shows the changes that occur in some common minerals at progressively higher shock pressures.

2. *High strain-rate effects* that can dislocate and destroy crystal lattices, form shock lamellae, and transform minerals into isotropic phase (e.g., feldspar to maskelynite).

3. *High temperature effects* are produced by the very high temperatures associated with shock-wave phenomena; for example, quartz may be fused to produce lechatelierite, or zircon may decompose to baddeleyite, at temperatures above 1500 °C. These effects can occur with non-impact processes, such as lightning strikes, but when considered with other factors can aid in the identification of impact events.

When the first lunar samples were returned to Earth and examined, there were abundant examples of shock metamorphic features, including fracturing, granulation, shock-lamellae and fused glasses. Table 8.1 classifies some of the shock metamorphic effects observed in the Apollo 11 rocks. These and other features have been observed in lunar samples from all sites. The breccias show by far the greatest shock-metamorphic effects, and as one would expect, the abundance and variety of shock metamorphic features increases with the apparent age of each site.

The recognition of shock metamorphic effects was not only important for substantiation of impact cratering on the Moon but has become an important means of recognizing impact events on Earth.

8.4.   *Regolith breccias*

Breccias in the regolith consist of rock fragments welded together in a finer matrix. Mare regolith consists of both breccias and bedrock lavas; conversely in highland regions, where the bedrock is itself breccia,

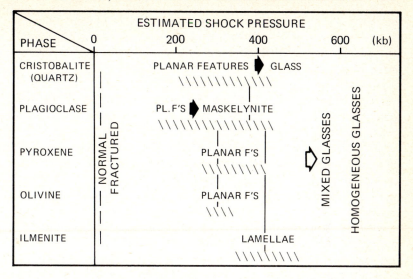

Fig. 8.9.   A scale of shock metamorphism (with estimated peak pressures) applied to the principal minerals in Apollo 11 rocks, together with various characteristic shock features formed over the pressure intervals indicated. (From Dence *et al.*, 1970.)

the regolith consist of many different types of breccia (see Chapter 9).

Breccias associated with break-up of local bedrock rather than being re-excavated breccia bedrock have been found at all lunar landing sites and are called *regolith breccias* (sometimes called soil

Table 8.1.   Degrees of shock metamorphism in lunar samples. (From Levinson and Taylor, 1973, after Chao.)

Degree of shock	Description	Pressure	Temperature
Weak	No intragranular deformation in plagioclase; sparse mechanical twins in ilmenite and pyroxene; microfractures in all minerals	150 kilobar (15 MPa)	—
Moderate	Mechanical twins, deformation and shock lamellae in plagioclase; abundant mechanical twins in ilmenite and pyroxene; abundant microfractures	250 kilobar (25 MPa)	?
Strong	Partial or complete vitrification of plagioclase; intense fracturing and mechanical twinning in pyroxene and ilmenite; etc.	500 kilobar (50 MPa)	?
Very strong	Flow in feldspar glass	?	700°C
Intense	Melting of entire rock by shock-induced heating; etc.	?	1200°C

breccias). These breccias are not related to single impacts but are consolidated debris that has been worked by numerous impacts before lithification. The clasts tend to be smaller than those of highland bedrock breccias because of repeated break-up, and only a few per cent of the clasts are larger than about 1 mm. Although many clasts may be recognized as local bedrock there are also many glass chips and spheres and much of the matrix is also made up of glass.

A complex history for regolith breccias is indicated by the presence of breccia clasts within the breccias. Chemical analyses show that there is a relatively high content of iron and of siderophile elements (those having an affinity with iron) and volatile elements normally associated with meteorites, as expected in an impact-generated rock unit of this type.

Regolith breccias and the enclosing unconsolidated fines are the only rocks that represent the Moon's surface history for the past 3000 million years. That only a few metres of rock should have been formed on the Moon for more than half its lifetime indicates the relative 'lifelessness' of this planetary body.

9. The highlands (terrae)

9.1. *Introduction*

The highlands (fig. 1.3) have received relatively little attention geologically compared with the maria. At first sight this is surprising since they make up most of the farside of the Moon and more than 50 per cent of the nearside. Volumetrically they are much more important than the maria: whereas the maria consist of a relatively thin skin only a few kilometres thick, the highlands represent the top surface of a crust tens of kilometres thick. The reason for a lack of attention to the highlands is essentially that the highlands are very much more difficult to interpret than the maria and younger basin deposits.

The highlands are densely cratered and although it is possible to obtain relative ages of individual craters by their state of degradation it is difficult to establish wide-scale stratigraphic horizons. Moreover, areas between craters are relatively nondescript, having few features that can be used to indicate the origin of the underlying units.

Samples and high resolution Apollo photography have done much to improve our knowledge of the highlands, leading to better known stratigraphy and origins for the units. Nonetheless, there are still many fundamental problems to be solved, two of the most important of which are first the role of volcanism in highland development, and second the identification of units associated with ejecta from the multi-ringed basins. A particular enigma is the origin of the smooth plains often called the Cayley plains (or Cayley Formation for the rock unit); these have been interpreted in various ways and were the materials on which Apollo 16 landed.

9.2. *Crater chronology*

Because cratering, the main process degrading the lunar surface, is essentially uniform as a function of time over the whole lunar surface, the degree of degradation of a particular crater will be indicative of its relative age (fig. 9.1). Thus, any two craters of similar size that show similar degrees of degradation may be designated similar ages. Careful

Fig. 9.1.   View across the highlands showing large degraded crater rims pock-marked by smaller craters. Small fresh craters are bowl-shaped. In the foreground is a floor-fractured crater. (NASA Apollo 17–0837.)

study of craters indicates a continuum of morphological stages indicative of degree of impact degradation. Pohn and Offield (1970) have proposed criteria to distinguish craters of different ages based on state of degradation, and used this method to provide relative dates for both individual craters and surfaces on which the craters are superimposed. The method assumes that all fresh impact craters in a given size range had the same initial morphology and that departures from the assumed original morphology are caused only by degradational ageing. Some confirmation of the method of Pohn and Offield is given by the observed superposition relations between craters where fresher craters are adjacent to more degraded ones.

Use of degree of crater degradation is generally applicable to dating lunar craters but there are several considerations that may affect age relations established by the method. Although the craters become modified mainly from meteoritic impact erosion and subsequent mass-wasting, there are other processes that may modify them, including structural modifications, volcanic process and intense but local moonquake activity causing significant shaking down of surface material. Lithologies of the rocks may also be different in craters from one area to another causing differential erosion. In addition the impact of a large crater will locally degrade those craters near to it by intense bombardment of secondary materials at high velocities, again making the affected craters appear older than those of the same age further away. This is known as proximity ageing.

The method of relative dating of craters by their degree of degradation depends on the assumption that craters always have the same initial form, an assumption which is not always true. The classic example is in the Orientale region where there are two craters Kopff and Maunder that post-date Orientale ejecta and therefore are both relatively young; yet Kopff does not have the normal ejecta facies around it nor the well defined secondary crater fields (Chapter 6.6 and fig. 3.10). Whatever the cause of this difference between two craters of similar age, it is clear that not all lunar craters start with the same morphology and attempts to date craters in terms of their eroded morphology may well, in some cases, give anomalous results.

Large craters appear to degrade in the following way. If we start with a fresh crater (with rays, radial and hummocky ejecta facies, secondary crater fields, sharp rim features and inner terracing), then the first stage in erosion will be the removal of the thin ray material and softening of features at metre resolution. At this stage of degradation of the larger craters, small craters (<20 km) will have virtually lost the surface patterns on the ejecta sheet and the crater rim will have become subdued. Continued degradation of large craters causes more subduing of form with a noticeable population of superimposed craters less than about 3 km across; the radial facies becomes almost indiscernible. Still older craters will have lost their secondary crater fields and the inner walls become more subdued, although terracing remains clear. Moreover there is usually at least one superimposed crater larger than about 3 km across. With increasing age the number of superimposed craters becomes greater and eventually the only evidence of an existing crater is a battered rim consisting of a circular cratered ridge. Much of the degradation of craters occurred before the end of heavy bombardment; after this the impact rate was too low to greatly modify major surface features.

Although the analysis of crater degradation is important and to some extent allows large structures such as the older basins to be relatively dated it does not provide the basis for widescale highland stratigraphy. An understanding of basin geology is of greater importance to understanding the evolution of the highlands. Lunar landings and returned samples have provided a new insight to this problem.

9.3. *The Apollo 16 landing site*

The Apollo 16 landing site (fig. 9.2) in the highlands was chosen to allow examination of highland inter-crater units then thought by many to consist of highland volcanic rocks. The landing site lay to the northwest of the Nectaris Basin and north to the crater Descartes on an area of smooth highland light plains, typical of Cayley Formation, adjacent to a more rugged ridged and grooved unit known as the Descartes highlands unit. It was the aim of the mission to sample both these units which could be taken as representative of extensive outcrops of units with similar morphology elsewhere in the highlands.

The upper surface of the Cayley Formation at the Apollo 16 site is smooth but broadly undulating with a maximum relief of several metres. Samples were taken by Apollo 16 astronauts Young and Duke at nine different stations on a 7 km traverse with the aim of collecting from different depths in the unit. This was accomplished by sampling on the rims of craters of different diameters each of which would have excavated material from different depths. Cayley material was found to consist of light and dark breccias containing fragments of plutonic anorthosite and feldspathic gabbro. It appears that the Cayley unit is crudely bedded with alternating layers of light and dark breccias, and estimates of depth of excavation imply that these lithologies extend down to at least 200 and possibly 300 metres below the surface (Muehlberger *et al.*, 1972). The breccias have been shock-metamorphosed and are interpreted as impact breccias. However, they do not show multiple brecciation to the extent of rocks from the Fra Mauro Formation sampled at the Apollo 14 site.

The Descartes highlands unit was sampled on the lower flanks of Stone Mountain. This more rugged unit rises 514 metres above the level of the Cayley plains and is strongly ridged and furrowed with numerous superimposed elliptical craters. Despite the strong difference in morphological expression of these two units, sampling of the Descartes highlands showed the rock types at least at the sample sites

Fig. 9.2.   Apollo 16 landing site (X) near Descartes. Smooth Cayley plains are marked (C) and Descartes highland unit (D). Picture is about 150 km across. (NASA Lunar Orbiter V 89 H$_3$.) Irregular patches on picture are processing defects introduced on the spacecraft.

to be identical in character to those collected from the Cayley unit, suggesting that the origins of these two units are closely linked; but despite the availability of samples, the origin of the units sampled is still under debate.

9.4. *Descartes highland type units*

These hilly, and ridged and furrowed units are not as extensive as the Cayley plains (fig. 9.3) and although they are found in many places in the highlands, they are particularly well developed on the west and north-west flanks of the Nectaris Basin (near the Apollo 16 site), and also on the west flank of the Humorum Basin. Trask and McCauley (1972) suggest as possible origins for these units that they are: (1) ejecta from large basins, or (2) highland volcanic rocks. These authors point out the morphological similarities between these units and other units surrounding Orientale interpreted as basin ejecta; but on the other hand note the presence of elliptical craters on the ridges and suggest that these are volcanic craters on dome-like volcanic structures. If these units are younger than Imbrium, then they can hardly be ejecta from other basins such as Nectaris known to be older than Imbrium.

Nevertheless, on the basis of the Apollo 16 samples, it is difficult to maintain a volcanic hypothesis for the Descartes highland type of unit, and the strong morphological similarity between these units and other basin ejecta units even without the evidence of samples would still argue strongly against a volcanic origin. It appears that a better interpretation of the units is that they are basin ejecta which have elliptical secondary craters produced by the Imbrium event superimposed on them; in this case they must be older than Imbrium and could be ejecta from basins such as Nectaris and Humorum.

Because craters are the dominant highland landform, highland stratigraphy consists principally of a sequence of overlapping units of crater and basin ejecta. This concept was accepted before the Apollo sampling but the difficulty had been to recognize individual rock units associated with the older basins except for the ring mountains and Fra Mauro type units associated with Humorum and Nectaris. The recognition of the ridged and furrowed terrain as basin ejecta provides a basis for unravelling highland stratigraphy. It is not always possible to determine the parental basins for individual units, but a first attempt at delimiting basin provinces is given in fig. 3.3.

Studies of the Nectaris area have allowed Stuart-Alexander and Wilhelms (1974) to define a new stratigraphic system representing the time between the Nectaris and Imbrium events (Chapter 10). If a stratigraphic column is considered to consist of a number of defined

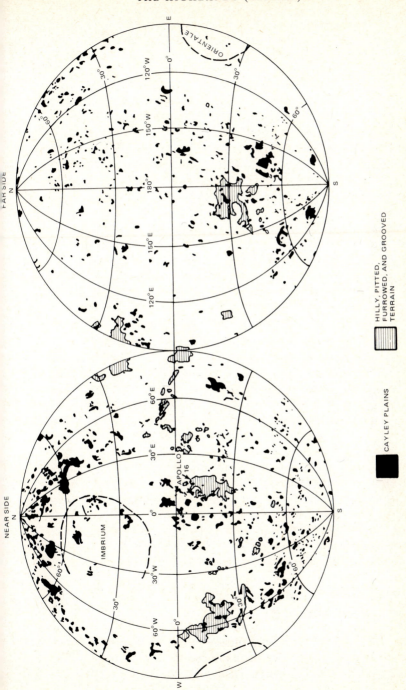

Fig. 9.3.   Distribution of Cayley plains and other units associated with impact basins. (After Howard *et al.*, 1974.)

extensive isolated events, then these events split up the history of the planet into identifiable time periods. Considered in this way the lunar highlands are less difficult to understand than previously.

9.5. *Cayley Formation*

These plain units (fig. 9.3) are the most problematical units of the highlands. They have smooth level upper surfaces (fig. 9.4) with an albedo similar to that of the surrounding highlands. Crater counts and overlapping relations indicate that they are older than mare units. Based on photogeology there are at least three explanations of the plains:

(1) That they are volcanic rocks produced either as pyroclastics, or as the highland equivalent of the maria flood volcanism but of more feldspathic lava than the maria. On this hypothesis, the breccias collected at the Apollo 16 site would be interpreted as the impact-brecciated top surface of a lava.

(2) That they are impact breccias from the large basins. If this explanation is correct, the most likely mode of implacement is by a kind of flow (possibly in a fluidized state) different from that of the denser fragmental flows which produced ridged and hilly terrain.

(3) That the material is local bed-rock that has been churned by dense secondary cratering from basins and large craters.

It is difficult to determine which of these mechanisms, or variations and combinations of them, was responsible for the Cayley plains. It is clear from the samples that if the Cayley plains are flood-lavas they have been extensively modified by impact to a considerable depth. If pyroclastic, they are quite different from any known terrestrial rocks produced by volcanic explosion. If they are volcanic, the complete lack of hydrous minerals in them leads to the question of what was the driving force of the explosions, which might also be asked with regard to Apollo 17 pyroclastics. The answer may be that volatile substances other than water have produced explosions. Certainly the volcanic hypothesis is the simplest based on photogeology, but it is the least likely based on studies of the samples. Explanations that relate the plains to large impact events suffer from the present lack of understanding of such major shock-wave events and the nature of the rocks produced.

In early lunar mapping, all highland plains were included in the Fra Mauro Formation, into which they were considered to grade. Later, the smooth plains were distinguished from the Fra Mauro and given the name Cayley Formation. Eggleton and Schaber (1972) note that in some places on the Cayley there are sinuous scarps similar to those found in the Orientale ejecta. These and other arguments have been

Fig. 9.4. Typical plains of Cayley Formation inside Ptolemaeus (about 150 km across).
(NASA Apollo 16–0579.)

used to indicate that Cayley material was transported as a fluid system
to greater distances from Fra Mauro at the end of the Imbrium event.
The material pooled in low places and would have consisted of solid
fragments together with drops and shards of rock melt which on
emplacement formed the breccias consisting of fragments in a glassy
matrix such as those collected at the Apollo 16 site.

That the Cayley Formation is churned-up bedrock has been put
forward by Oberbeck *et al.* (1973b) who argue that at great distances
from basins and large craters the velocity of material on ballistic
trajectories was so high that the impacting debris excavated considerably more than its own volume of material in secondary craters (fig.

5.5). They consider that the highland surface has been broken up by secondaries to form large volumes of brecciated material. This material then moved down slope as part of the impact process as well as being shaken down by seismic activity to fill topographic depressions. Some support for this is given from Apollo photographs which when viewed stereoscopically show that not all the Cayley areas are as flat as was once supposed but slope up towards the surrounding higher ground. Head (1974) considers that old craters near to Cayley plains were the primary contributors of material to this unit.

The origin of the Cayley materials is far from being solved to everyone's satisfaction. It is likely that, for the Moon as a whole, these materials represent more than one event and that Cayley plains may have more than one origin. It is dangerous to extend the conclusions from one landing site to all the highland plains units, and although the impact process has been emphasized here, the possibility that the plains were produced by some form of explosive volcanism unknown on Earth still remains at least in the minds of some workers.

9.6.   Highland breccias

Rocks returned from all the landing sites show that except in the maria, breccias are the dominant near-surface bedrocks on the Moon. The highland sites visited appear to be underlain by breccias of different types, and the presence of various degrees of shock metamorphism indicates to most workers that they formed as a result of rock fragmentation, melting, and lithification by impact process (James, 1974). Many have textures that are similar to those of breccias from terrestrial impact craters. Similarities between breccias at the Descartes site (Apollo 16) and other sites more obviously associated with impact basin ejecta (Apollo 14 and 15, Imbrium; and Apollo 17, Serenitatis) suggest that most if not all highland bedrock breccias were formed by major impact events.

At the Apollo 16 site two types of breccia are important. The major type is known as *cataclastic anorthosite breccia* consisting of coarse-grained plutonic rocks that have been crushed, or both crushed and recrystallized; some show evidence that this has occurred several times. These breccias are white and are texturally similar to breccias in the walls of terrestrial impact craters, and some types of ejecta in which shear and granulation dominate over shock-metamorphic events. The petrology of the clasts suggest that they crystallized from a magma deep in the Moon's crust before being incorporated in the breccias.

The other major type of breccia at the Apollo 16 site is a *black and white breccia*; the white part is similar to the cataclastic anorthositic breccias while the black material consists of melt intruded into the white material. This black intruding material was molten on emplacement and contains sparse clasts of white rock together with nickel–iron globules. These breccias show complex injection and deformation sequences and it appears that the black veins were injected while the white rocks were undergoing cataclastic deformation. The mobile injections solidified rapidly and then deformed by continued fragmentation. Such rocks are interpreted as crushed bedrock intruded by impact melt in the early stages of crater excavation when the crater walls were subjected to high temperature and deformation. These samples support an impact origin for Cayley plains, but the mechanism of emplacement is still uncertain.

Similar rocks to those at the Apollo 16 site are found more closely associated with the basins. Black and white breccias occur widely and are an important rock type at sites close to circum-basin mountains (Apollo 15 and 17). One rock at Apollo 15 consists of cataclastic white rock injected by black, dense aphanitic material with abundant white clasts. Some of the clasts however are not local and some have a history of more intense shock than the host white rock.

The black component of these breccias also has similarities with the so-called *light grey breccias* found in the Taurus-Littrow site (Apollo 17) associated with impact debris from the Serenitatis basin. There are several different types of this breccia but generally they are fragment-laden devitrified glass similar to suevite (Chapter 5.3), aggregates of fragments bonded by devitrified glass, or less well consolidated aggregates of the previous two types of grey breccia. One sample collected is oval in shape with a vesicular outer rind and may be a throw-out 'bomb'. Differences in thermal history of the clasts suggest mixing of hot and cold fragments. Other rocks at the Apollo 17 site and in the Fra Mauro Formation are similar to the grey breccias. For example, *blue–grey breccias* differ from the grey breccias by having some interaction between the clasts and glassy materials.

The major rock type in the Fra Mauro Formation (Apollo 14) is that known as *thermally metamorphosed breccia*. A few of these breccias were returned from the Apollo 15 and 16 sites also. This type of breccia may represent lithified debris from the continuous ejecta of big basins, but the high degree of thermal metamorphism obscures other characteristics that might have been diagnostic of their origin (James, 1974). As pointed out in Chapter 3, it is possible that those from Fra Mauro were metamorphosed before they were incorporated in Imbrium's ejecta. This type of breccia differs from regolith breccias (Chapter 8) in having a less diverse suite of lithic fragments, but on the

Fig. 9.5. Large (~80 km diameter) crater north-west of Mare Ingenii on the lunar
farside showing extensive grooves on the crater wall and rim, and hummocky
terrain around the crater. This region is antipodal to the Imbrium basin and the
unusual terrain may have resulted from focused seismic energy generated by the
Imbrium event. (NASA Apollo 15–0084.)

other hand they are not as homogeneous as the Apollo 17 light grey
breccias. In the fines (0·1–1 mm size range) fragments of single mineral
grains are twice as abundant as in regolith breccias.

On the rim of one crater at the Apollo 14 site there were also white
boulders. These were probably derived from deep in the Fra Mauro
Formation. The rock type is known as *glass-poor feldspathic breccia*,
consisting of friable crushed debris from highly feldspathic rock.

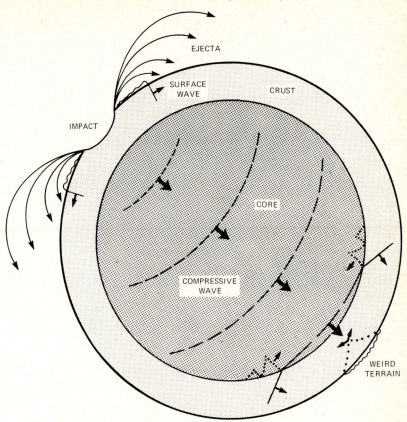

Fig. 9.6. Hypothetical diagram illustrating the possible focusing of seismic energy on the side of Mercury antipodal to a big basin-forming impact. Similar effects are suggested for the Moon. (Courtesy P. Schultz.)

9.6. 'Weird' terrain

A rather unusual terrain (fig. 9.5) in which the cratered surface is heavily fluted and pitted occurs in two areas of the highlands, one of which is antipodal to Imbrium and the other at the antipode of Orientale. In these areas there are grooves down the inner walls and on the outer rims of large craters and other inclined surfaces. Many of the smaller craters are furrowed and have relatively smooth walls without the usual inner terracing. The age of these terrains is pre-mare and is probably similar to the age of the basins.

Two suggestions have been put forward to explain this 'weird' terrain (so named informally by the Mariner 10 team when a similar type of terrain was observed on Mercury antipodal to the Caloris

Fig. 9.7.   The 110 km diameter floor-fractured crater Gassendi on the north edge of
Mare Humorum. (NASA Lunar Orbiter V 178 M.)

impact basin). Moore *et al.* (1974) suggested that there is a concentration
of ejecta in the antipodal regions to a large basin impact and thus
explain the weird terrain as being furrowed and pitted by high velocity
secondary impacts, Schultz and Gault (1975) also explain the terrain in
terms of Imbrium and Orientale impact but suggest that the terrain
results from extensive mass-wasting caused by seismic effects enhanced
at the antipodes of the basins (fig. 9.6). They show that vertical surface
displacements from surface waves at the antipode could have been of
the order of 10 m. Reflected waves will also converge within the Moon
beneath the basin antipode to produce deep-seated crustal weaknesses.

9.7. Floor-fractured craters

This term was coined by Schultz (1976) for those large craters (mainly occurring in the highlands) that have concentric and radial patterns of rilles on their floors (figs. 9.7 and 11.10). Many of these craters have been modified by emplacement of mare-like plains before cracking occurred. They tend to concentrate close to the margins of mare materials but a substantial number occur in the farside highland regions. Maximum mean distance from the nearest mare is 300 km. From a detailed study of these craters Shultz has shown that the central peaks of floor-fractured craters stand relatively high compared with those of normal lunar impact craters. He therefore argues that the floors of these craters has been pushed up, presumably by injection of magma into the brecciated region below the craters (fig. 9.8). Normal impact craters can, apparently, be modified in several different ways indicating variations in the sequence of magma emplacement and extrusion.

From our previous discussion of the maria it seems likely that mare emplacement was accompanied by the injection of extensive sills into the mare regions and underneath the surrounding highlands. It is thus possible that in the floor-fractured craters we are seeing areas where sills have intersected the brecciated and faulted regions below impact craters, and that sill material has intruded upwards below the floor of these craters to push up the floor. Clearly, further study of these craters could improve our understanding of the mechanism and style of igneous activity on the Moon.

9.8. Composition of highland crust

Samples from highland crust have been collected from a number of different places. Apollos 14, 15, 16 and 17 all sampled the highland crust either as material ejected from the Imbrium, Serenitatis and Nectaris basins or uplifted in the mountains surrounding the basins. The Soviet Luna 20 collected rocks in an area between Crisium and Fecunditatis. Further information on the composition of the highland region was obtained from X-ray fluorescence determinations made in lunar orbit, providing information on Si, Al, Mg and radioactive elements.

It is quite clear that the highlands differ in composition from marial rocks by being richer in alumina. Moreover, the highlands are not homogeneous and there are regional differences in Al/Si and Mg/Si ratios. A detailed understanding of the highland compositions is not easy because any original crustal features have been destroyed by

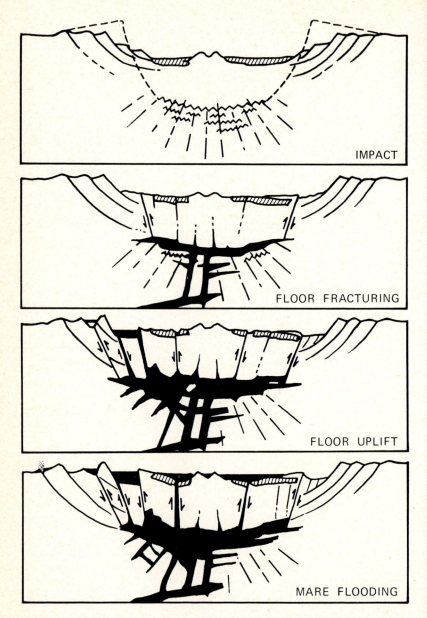

IMPACT

FLOOR FRACTURING

FLOOR UPLIFT

MARE FLOODING

Fig. 9.8. Hypothetical sequence (from top to bottom) of development of floor-fractured
        craters. The dash-line in the top figure shows the initial excavation cavity before
        adjustment of floor and wall collapse immediately after impact. (From Schultz,
        1976.)

numerous impacts giving a brecciated upper surface which, from seismic data, appears to be about 10 km thick. The seismic data also suggests that the crust is fractured to a depth of up to 25 km. Thus the textures in highland rocks reflect their impact nature and much of the evidence for the original pre-impact constitution of the crust has been destroyed.

Fragments of igneous rocks within the breccias include (table 9.1) anorthosites, gabbroic anorthosites, highland basalts, anorthositic norite and troctolite, low-potassium Fra Mauro basalt, medium-potassium Fra Mauro basalt (non-mare basalt) and also dunite at the Apollo 17 site. The textures of these rocks are variable and in some cases they may be impact-melts that have crystallized with igneous textures, but the degree of impact melting is unknown and provides yet another problem to the highland story. Certainly tens or even hundreds of cubic kilometres of melt may have been formed by major impacts (Head, 1974). Impacts might have produced fractional melting although it is much more likely that differences from the original composition were produced by normal differentiation in a large body of the melt. Loss of elements by differential vaporization may have taken place, but the quantity lost was probably low (Taylor, 1975).

Table 9.1.   Rock types and their synonyms for igneous clasts in highland breccias.

Rock name	Synonym	Main minerals	Al_2O_3 %
Anorthosite	—	Mainly Ca plagioclase	33–36
Gabbroic anorthosite	Noritic and troctolitic anorthosites	Plagioclase and ortho-pyroxene or olivine	28–33
Anorthositic gabbro or highland basalt	Anorthositic norite and troctolite	Plagioclase, orthopyroxene and olivine	24–28
Troctolite	Spinel and troctolite	Plagioclase and olivine with spinel	20–25
Low-K Fra Mauro basalt	High-Al basalt Low K. KREEP	Plagioclase, orthopyroxene, olivine, clinopyroxene and ilmenite. Composition low in K, rare earths, P, V, Th. etc.	17–23
Medium-K Fra Mauro basalt	KREEP, KREEP basalt Norite	Similar to low K Fra Mauro basalt but with relatively high K content ($\sim$0·5%)	15–20
High-K Fra Mauro basalt	Non-mare basalt	As above but with more than 1% K.	

Notes: 1. Some workers use the term ANT for the range of anorthosite–norite–troctolite rocks containing 28–35% Al_2O_3.
2. *Dunite* consisting of large pale green olivine crystals in a crushed olivine matrix was found at the Apollo 17 site. This rock yields an age of 4600 million years.

Like the rocks of the marial regions, those of the highlands are made of non-hydrous minerals except for one sample at the Apollo 16 site which is a shock-melted breccia containing small amounts of goethite ($FeO \cdot OH$). The origin of this hydrous mineral is unknown but it could have been produced by alteration of the rock on exposure to the Earth's wet atmosphere when the sample was brought back from the Moon. Apart from this one piece of ambiguous evidence, it is apparent that the highland rocks were all emplaced 'dry'.

The sample analyses provide confidence in the X-ray fluorescence and other orbital data. These data show some interesting variations, particularly a high gamma-ray count in the areas of Imbrium ejecta. There is a sharp compositional boundary between the highlands and the maria, confirming that impact mixing is not an efficient process.

There is still room for a wide range of ideas about the early stages in lunar history as represented by the highland crust. Two extreme ideas are: (1) that the highland crust is a chemically distinct refractory layer accreted or condensed during the last stages of planetary formation, or (2) that the alumina-rich crust is derived by differentiation of the whole Moon during extensive melting early in lunar history. Study of the highlands could be directly relevant to a fundamental question as to whether the Moon was initially a homogeneous or heterogeneous body. The surface form of the present highlands indicates that impact cratering was the dominant process in the formation of the upper surface of the crust and it is not known what type of igneous activity was taking place during the first half billion years of the Moon's history. However, Wood and Head (1975) have suggested that spectrally distinct, pre-mare red spots may represent rocks from early volcanic activity.

9.9. *Age of the highland crust*

Samples do not provide unambiguous information about the age of the events in highland history. If we assume that the highlands are made up of an early igneous crust, there are two obvious questions to ask: (1) what are the ages of the large basins? and (2) what is the age of the igneous activity?

Ages of highland rocks cluster between 3900 and 4000 million years (Taylor, 1975). This date may represent a general age for the basins, or, alternatively, it may be a sampling effect because all landing sites were close to just a few of the total number of basins. Although it is clear that all the basins were formed before 3800 to 3900 million years ago it is not known whether the basins were formed at intervals after the Moon was formed, or whether they represent a discrete phase of

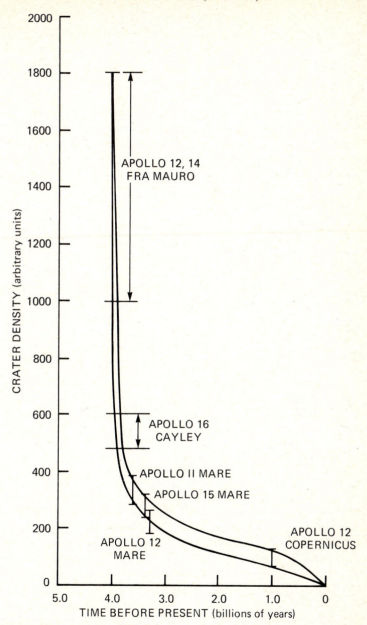

Fig. 9.9. Total accumulation of impact craters on the lunar surface as a function of age. The approximately linear increase in degree of cratering from the present back to about 3500 million years ago implies a fairly constant flux for that period. Before then there was a very high but decreasing flux. (From Soderblom *et al.*, 1974.)

cataclysmic bombardment near to 4000 million years ago. There has been much discussion about whether bombardment earlier than 3900 million years ago was a slowly diminishing process with Orientale being the last in a series of impact events, or whether this early bombardment was episodic, the present basins having been formed by a final episode of bombardment. The clustering of highland ages may suggest that the basins represent an episode in bombardment history but this conclusion is not definitive because sampling has not been wide enough.

Rubidium and lead isotope ages suggest that the original material that formed the crust is 4600 million years old. The date of an individual igneous episode may be given from a relict plagioclase crystal which appears to have crystallized over a period between 4500 and 4250 million years ago.

In summary it appears that igneous activity occurred at least between 4.5 to 4.24×10^9 years ago. Certainly there was formation of big basins between about 4 and 3.9×10^9 years ago and it is not known whether there are basins older than this.

What was happening on Earth during the early lunar period? This is not an easy question to answer because the terrestrial geological record becomes increasingly difficult to interpret before about 600 million years ago. The oldest dated rocks on Earth are nearly 4000 million years old, shortly after the time of the end of heavy bombardment on the Moon. Geological studies suggest that at that time conditions on Earth were different from those pertaining at the present time. The Earth's atmosphere was probably a reducing one (hydrogen and hydrogen-containing gases or carbon dioxide) until 2000 million years ago when simple life-forms started to change it to an oxidizing one (containing oxygen, as now).

Evidence from the very early pre-Cambrian rocks shows that the development of a crust on Earth goes back to at least nearly 4000 million years ago (Read and Watson, 1975) and, at the time when the Orientale and Imbrium basins were formed, on Earth typical continental crustal processes were occurring including granite formation, dyke intrusion, volcanism and sedimentation. Of particular interest is the presence of large amounts of anorthositic rocks in the early crust of the Earth. These are well exposed in Greenland where they occur as 3500 millions years old thick bands up to 1 km thick, often distorted and broken by tectonism.

There is some evidence that the Earth's tectonic regime was different from the present in these early phases of its history. The presence of stable crustal areas separated by mobile belts typical of the more recent history of the Earth appears to go back to about 2800 million years ago.

At that time many small cratons (stable continental masses) were probably enclosed by a network of mobile belts, but with time the mobile belts became progressively stabilized giving rise to much larger stable masses. This period of building crustal plates may also be associated with two, apparently world-wide, events at 2300 million years and about 1200 million years consisting of massive intrusions of basic materials.

Plate tectonics on Earth as we know it today may not have developed until about the time of the last marial outpouring on the Moon. There was, however, a continental crust of granitic and anorthositic material together with vast outpourings of basic lavas represented by green-stone belts consisting of thick sequences of basaltic material overlain by intermediate and acid composition lavas. These are now seen squeezed between granitic crustal masses.

From our present knowledge of other planets it seems unlikely that the Earth should have avoided the period of bombardment that was widespread in the solar system and it is thus likely that the Earth's crust was modified strongly by impact craters of varying sizes—at least in the areas not covered by the oceans. There is still much to learn about these early phases in the Earth's history which appears to be so different from present day conditions, and a comparative study between the Earth and the geological history as revealed by the Moon and other terrestrial planets is important to our understanding of planetary evolution.

Table 9.2.   Typical analysis of igneous rocks found as clasts in highland breccias. (From Taylor, 1975.)

	Anorthosite	Gabbroic anorthosite	Anorthositic gabbro	Troctolite	Low-K Fra Mauro basalt	Medium-K Fra Mauro basalt
SiO_2	44·3	44·5	44·5	43·7	46·6	48·0
TiO_2	0·06	0·35	0·39	0·17	1·25	2·1
Al_2O_3	35·1	31·0	26·0	22·7	18·8	17·6
FeO	0·67	3·46	5·77	4·9	9·7	10·9
MnO	—	—	—	0·07	—	—
MgO	0·80	3·38	8·05	14·7	11·0	8·70
CaO	18·7	17·3	14·9	13·1	11·6	10·7
Na_2O	0·80	0·12	0·25	0·39	0·37	0·70
K_2O	—	—	—	—	0·12	0·54
Cr_2O_3	0·02	0·04	0·06	0·09	0·26	0·18
Total	100·5	100·2	99·9	99·9	99·6	99·4

10. Stratigraphy

10.1. *Introduction*

Stratigraphy is that branch of geology dealing with the description, distribution, classification and chronological succession of layered rock units. The aim of stratigraphy is to reduce the enormous complexity of a planet's surface and near-surface rocks to a level of understanding that will allow the geological history to be unravelled, a goal which is achieved with the assistance of geological mapping, the 'hand-maiden' of stratigraphy. Geological maps are graphic presentations that link observations made at different localities into a unified, comprehensible form. When properly derived, geological maps display the three-dimensional shapes of rock units and their geometrical relations to one another. Because rocks are the very heart of geology, geological maps provide a framework for nearly all geological investigations. In this chapter we shall consider the rationale of stratigraphy in general, lunar stratigraphy in particular, and geological mapping; we then discuss the lunar geological column as it is presently understood. An excellent discussion of lunar geological mapping is given by Wilhelms (1972).

10.2. *Geological mapping*

Throughout the history of geology, a recurring problem has been to understand the relation between units of rock and sub-divisions of geological time represented by the rocks. Three categories of geological units are defined (see American Commission of Stratigraphic Nomenclature, 1970).

(1) *Rock-stratigraphic units*, the basic units of stratigraphic mapping, are sub-divisions of the crust based primarily on lithological attributes and geographical continuity. Rock-stratigraphic units in decreasing magnitude are *group*, *formation* and *member*, of which the formation is the fundamental unit consisting of a body of rock characterized by lithological homogeneity. Rock-stratigraphic units do not *necessarily* correspond to units of time. For example, a rock unit may, from bottom

to top, cross an important time boundary established by rock sequences elsewhere because it was in continuous processes of formation at the time that breaks were occurring in sequences in other places. Alternatively, the time span of a rock unit may differ from one area to another.

Rock-stratigraphic units consist of materials that result from specific geological processes. For example, a volcanic eruption produces a lava flow which crystallizes into a three-dimensional rock body with a specific lithology, covering a specific span of geological time. That same unit can later be weathered into particles, eroded, and the particles deposited to form another three-dimensional body which also has a specific lithology and represents a specific span of geological time at one place, but may represent different age spans in different places. It is rock units that are recorded on geological maps, these being the units that can be observed: other units involve a time interpretation. When rock units are placed in a sequencial order they give a *stratigraphy*.

(2) *Time-stratigraphic units* are rock units that represent specific intervals of geological time, independent of lithology or mode of origin. The magnitude of the unit is measured by the length of time the unit represents, not by the thickness of the rock unit. In decreasing magnitude, time-stratigraphic units are *Systems*, *Series* and *Stages*, the basic unit being a System.

(3) *Geologic-time units*, are units of time rather than rock units and are defined by time-stratigraphic units: geologic-time units in decreasing magnitude are *Eon*, *Era*, *Period*, *Epoch* and *Age*, where a Period corresponds to a System in Time-stratigraphic units.

A hypothetical planet formed from a molten blob into a completely homogeneous sphere completely lacking any modifying processes would be easy to map, as it would consist of a single rock-stratigraphic unit: but the Moon and other terrestrial planets are not like this, consisting instead of many rock units of different ages and origins (fig. 10.1). The complexity of a planet's stratigraphy seems to be directly related to the number and intensity of the geological processes that have operated during the history of the planet. We will discuss in Chapter 12 the differences between the inner planets in terms of their processes; it is fortunate that the first extraterrestrial planet to be geologically mapped was the Moon. We were able to go from the Earth, which appears to be the most complex of the inner planets, and for which we have the most information, to the Moon which appears to be simple in terms of geological processes and the resulting rock units.

Stratigraphic practices are derived from terrestrial geology, and are based on well tested geological principles: they should therefore be

Fig. 10.1.   Oblique view south over Guericke south of Imbrium. This shows the stratigraphical relations between a pre-Imbrium crater (cut by Imbrium sculpture),
plains in the floor of Guericke and mare material cross-cutting both these earlier
units. Elliptical craters to lower left of Guericke are partially buried secondary
craters possibly from Imbrium. (NASA Apollo 16–2486.)

applicable also to extraterrestrial geology. On a practical level, however,
there are some differences between terrestrial and extraterrestrial
mapping procedures. Whereas rock units on Earth have three-
dimensional forms that are fairly well-known, based, for example, on
exposed sections and drill-hole data, lunar units are known mostly by

two dimensions gained from photographs; their third dimension is inferred. In addition, terrestrial rock units are defined by their lithology; only where samples have been collected is the lithology of lunar units well established. For these and other reasons the term *lunar material unit* is used as the basic mapping unit on the Moon instead of the terrestrial *rock-stratigraphic unit.*

The first step towards establishing a stratigraphic column is to map the geology, and the first step of geological mapping is the recognition of units. Lunar material units are differentiated mainly on the basis of their surface characteristics, topographic expression and texture (presence or absence of ridges, hillocks, lobes, pits, etc.), the scale of the texture, albedo, colour differences and continuity of morphological 'sameness' (for example, areas in which all features are sharp and well-defined, in contrast to areas having features that are smooth and rounded). These photogeological techniques are supplemented by studies of radar surface properties, infrared emission properties, microwave properties and other remote-sensing methods, each of which contributes further information, but despite the sophistication of various methods, photogeology remains the primary means of making geological maps of planetary surfaces.

A recurring problem in photogeological mapping is the failure to distinguish between *geomorphological maps*, which show the distribution of landforms, and *geological maps*, which show the distribution of three-dimensional rock units. Nevertheless, there is a relation between surface topography and rock units. Topographical characteristics are used more than any other property to define units, on the basis that morphology is the surface expression of the manifold properties of the rock unit upon which the surface developed. However, identical morphologies can result from quite different processes, involving totally different rock units, for example smooth plains can be formed either by lava flows or by emplacement of impact ejecta. We attempt to map three-dimensional materials, not physiographic forms: thus we make lunar stratigraphic maps that show crater rim *material*, but not the crater, which is just a hole. The best test to determine if a mapped unit is valid is to draw a cross-section through it. Material units must show as three dimensional bodies which either overlie or are overlain by other material units.

Another problem common in planetary stratigraphy is the imposition of interpretative bias. Many lunar 'geological' maps have been made in support of topical studies and rely on the interpretations of the investigators. Thus, the maps produced by workers who believed in volcanic origins for craters tend to show volcanic rock units originating from craters and relations between the units to fit the interpretation. Other workers have produced maps which show rocks fitting structural origins for various features. To avoid this bias the lunar mapper must define material units objectively, that is, based on physical character- istics which can be seen by others. The convention in use since the initiation of the U.S. Geological Survey's lunar geological mapping programme and the application of rigorous stratigraphic principles in the 1960's is to separate the *descriptions* of lunar material units from the *interpretations* of their origin. If this convention is followed properly, a lunar geological map will remain valid because it will show the distribution and relative ages for lunar materials regardless of how many interpretations might be applied to the origin and lithology of the units. Many lunar geological maps have now been completed by the U.S. Geological Survey and a similar programme continues in the mapping of Mars and Mercury.

Once the units are recognized, defined by their properties and mapped for their areal distribution, the next step is to determine their relative age in relation to other material units. This is a procedure based on principles of sequence: younger units overlie or embay older units (fig. 10.1); the contact of a younger unit truncates the contact between older units. Copernicus (figs. 2.2 and 2.3) is clearly younger than the crater Eratosthenes and the surrounding mare material because its rim material, rays and secondary craters are superimposed on Eratosthenes and the mare material. The mare material, in turn, is younger than the crater Archimedes because mare material fills and embays Archimedes and truncates the contact between the Archimedes rim material and the surrounding plains.

10.3. *The lunar stratigraphic column*

In previous chapters we have shown that the basic structure of the lunar surface is related to the large circular basins. From the principle of sequence, most of the basins can be dated relative to each other (figs. 1.5 and 3.3). Thus, we have the beginning of a basis for determining stratigraphical succession.

Just as in terrestrial stratigraphy, lunar geological mappers seek evidence in the rock record for the same geological event over a wide area. Terrestrially, rock units useful in sub-dividing the stratigraphic column may include widespread ash layers and beds containing index fossils. On the Moon, ejecta deposits from a single large impact event are formed during a short interval of time; thus, the ejecta from each of the basin-forming impacts provide stratigraphic reference planes that are relatively widespread and which represent short intervals of geological time. Of course, the younger the basin, the better exposed and definable the unit. One of the best datum units on the Moon is the Fra Mauro Formation (fig. 3.9), consisting of ejecta deposits from the Imbrium basin. This unit is defined as marking the base of the *Imbrian System*. Apollo 14 was selected to land on the Fra Mauro Formation, and isotopic dating of returned samples places the age of the Formation and the impact that formed the Imbrium basin at about 3900 million years ago.

While the base of the Imbrian System is well defined by the Fra Mauro Formation, the top of the system is defined with great difficulty. Initially it was taken as the base of the mare materials, which themselves were included in a younger system, the *Procellarian System*. This system was discarded however, because neither the bottom nor the top of the rocks included in it could be considered the same age over the entire Moon. The Procellarian System was changed to a group of rock units that occur in the upper part of the Imbrian System (fig. 1.5). This still does not define the top of the Imbrian System, a problem that remains because a good datum plane is lacking.

The next youngest time-stratigraphic unit is the *Eratosthenian System*. Its base is characterized by the oldest craters larger than about 40 km in diameter that are essentially fresh—retaining most of their original features—but have lost their bright rays through degradation (fig. 2.2). Thus, some mare basalts are Imbrian while those superimposed on Eratosthenian-age craters are Eratosthenian or conceivably younger.

The youngest system on the Moon, the *Copernican System*, includes all materials of the bright-rayed craters.

All rocks and events older than Imbrian, including those associated with the basins that pre-date Imbrium, were lumped as pre-Imbrian (a term somewhat analogous to the terrestrial pre-Cambrian) until recently. Just like its terrestrial counterpart, the pre-Imbrian was a difficult unit to sub-divide; however, Stuart-Alexander and Wilhelms (1974) have now recognized the ejecta of the Nectaris basin (the Janssen Formation) and defined the *Nectarian System* immediately before the Imbrian. All rocks older than Nectaris are now referred to as pre-Nectarian.

10.4. *Evolution of the lunar surface*

The systematic geological mapping that has been completed so far for the Moon, combined with other lunar investigations, permits the assignment of relative ages for most large lunar features and the interpretation of the processes and events that lead to their formation. An

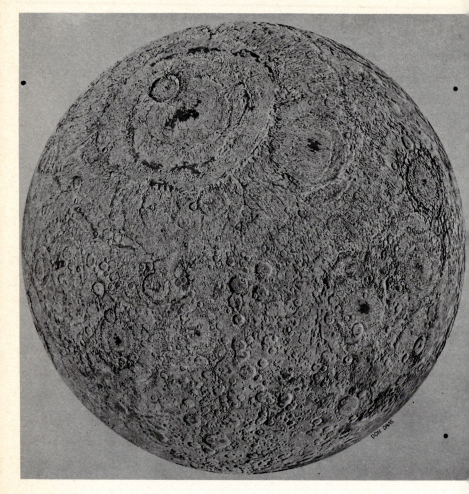

Fig. 10.2.   A reconstruction of the nearside of the Moon in mid-Imbrian times before
major mare flooding. (From Wilhelms and Davis, 1971.)

excellent graphical synthesis of the evolution of the nearside of the
Moon has been made by Wilhelms and Davis (1971), showing the
Moon at three different times in its history (figs. 10.2–10.4) based on a
combination of geological interpretation and artistic portrayal.

Figure 10.2 represents the Moon at about 3800 million years ago,
after the period of heavy bombardment and following the formation of
the multi-ringed basins. The mare flooding has just started. The
prominent basin shown is Imbrium, modelled after Orientale, with its
associated ejecta deposit, the Fra Mauro Formation. Note that some

Fig. 10.3. A reconstruction of the nearside of the Moon at the end of the Imbrian Period after emplacement of most of mare material at about 3300 million years ago. This is similar to the present view (fig. 10.4) except for absence of young craters such as Tycho, Copernicus and Eratosthenes. (From Wilhelms and Davis, 1971.)

craters, now barely discernible, are shown as fresh structures, such as the Iridum crater perched on one of the rings of Imbrium.

Figure 10.3 shows the Moon at about 3300 million years ago, after most of the mare material was emplaced, but before the formation of major post-mare deposits and craters. Lacking at this time are Copernicus and other bright-ray craters, the bright domes near Gruithuisen, and the mare domes and plateaus of the Marius and Rümker Hills.

Fig. 10.4.   The present Moon, to compare with figs. 10.2 and 10.3. (From Wilhelms and Davis, 1971.)

The present lunar surface is shown in fig. 10.4, which is prepared from Earth-based photography. Note the addition of bright-ray craters of Copernican age and the Eratosthenian craters. This series of diagrams summarizes the evolution of the lunar surface and the stratigraphic succession of units that make up the frontside of the Moon. Pre-Nectarian and Nectarian times are represented by a heavily cratered surface and the formation of the early multi-ringed basins. The start of the Imbrian Period is defined as the time of the impact that formed the Imbrium basin (fig. 10.2) and the base of the Imbrian System is the base of the Imbrium ejecta deposits, the Fra Mauro Formation. The rest of the Imbrian Period was marked as a time of emplacement of the thick flood-type basalts concentrated in the front-side basins (fig. 10.3) and also of diminished cratering. This was followed by the Eratosthenian Period, when the flood-type basalts gave way to less voluminous eruptions typified by the flows and flow units in the Euler region in Mare Imbrium. It is not presently known, however, if the transition from one type of basalt eruption to the other type occurred at the same time everywhere on the Moon—probably it did not. Copernican times are characterized by the formation of craters that are now seen as sharp rimmed, bright-rayed structures which reflect their comparative youth.

11. Internal structure and geophysics

11.1. *Introduction*

Before spacecraft ventured to the Moon, very little was known about the lunar interior or the nature of lunar tectonic processes. Photographs of the lunar surface showed suspected lava flows and faults which led to ideas on the existence of magma chambers and crustal deformation, but the ideas were highly speculative. Earth-based determinations of the lunar mass, gravity, orbital characteristics and gross magnetic properties permitted additional speculation on the lunar interior, beyond ideas that developed from the study of surface features.

Unmanned missions to the Moon, initiated by the Soviet *Luna* series in the late 1950's, and followed by the United States Explorer 35 Orbiter, returned geophysical data which began to place constraints on various models of the lunar interior. Subsequent missions culminated in the complex arrays of instruments placed on the lunar surface during the Apollo missions. The manned and unmanned lunar missions have produced a tremendous wealth of geophysical data, much of which has only begun to be examined. Lunar geophysics is considered here in terms of the lunar magnetic field, gravity variations, seismic activity, internal composition and structure.

11.2. *Magnetic field*

The Moon's magnetic properties have been determined by orbiting satellites, station magnetometers, portable magnetometers and laboratory measurements on returned samples.

The measurement of the lunar magnetic field began in January 1959, when Luna I came within several hundred km of the Moon. This mission was followed by Luna II and Luna X, which showed the existence of a weak lunar magnetosphere. In 1967 the U.S. Explorer spacecraft orbited the Moon with two very sensitive magnetometers, and results from this led C. Sonnet to conclude that, if a permanent lunar magnetic field existed at all, its magnitude would be less than

2 gamma (1 gamma $= 10^{-5}$ G $= 0 \cdot 1$ T) at an altitude of 830 km. The magnitude of the magnetic field at any one place is the sum of the external field (solar or terrestrial), the permanent lunar field, and the fields induced in the Moon by changes in the external field.

Measurements of the permanent magnetic field on the Moon range from 3 to a maximum of 313 gammas, an extremely small value compared with the Earth's field, which ranges from 30 000 to 60 000 gammas. Unlike rocks from Earth in which iron oxides are the main magnetic minerals, lunar rocks appear to have metallic iron as the dominant magnetic mineral. Most of this iron is in the form of fine particles produced during impacts. Maps of the magnetic field derived from satellite data show a distinct correlation with geological provinces, in which the highest fields occur over the cratered highlands and are especially associated with the Cayley Formation. Thus, the maximum field intensity measured on the ground is 313 gamma at the Apollo 16 site where impact-generated breccias are abundant, and the minimum intensity occurs at the Apollo 15 site, most of which is underlain by relatively undisturbed mare basalts. The individual samples have random magnetic orientations showing that they were magnetized before being thrown onto the surface by local impacts.

The origin of the magnetic field remains a problem. Three possible origins are currently under consideration:

(1) that there was an internally generated magnetic field similar to Earth's dynamo field but which no longer exists on the Moon,
(2) that there was a strong external magnetic source (solar or terrestrial) which imposed its field on the Moon, early in history,
(3) that the lunar field was produced from local sources, such as impact events.

All of these possibilities have problems, however. Lunar samples show that they were heated above their Curie temperature of 750 °C to 770 °C before 3000 million years ago, long after the departure of any magnetic field associated with the origin of the solar system. Although seismic evidence indicates the possible existence of a very small molten core, the core probably was never large enough to produce a 'dynamo' magnetic field of sufficient intensity to be imposed on the rocks. If, however, the Moon did have an early magnetic field then high remnant magnetism in Cayley breccias may be interpreted as resulting from impact melt cooling from above the Curie temperature in a magnetic field.

Local sources of magnetism, such as impact-induced fields, are considered by some investigators to be plausible, especially in view of results from the hand-held magnetometer at Apollo 16 and a correlation of intensity with crater frequency, but so little is known about shock-induced magnetism that the question remains open.

11.3. *Lunar gravity data*

The external gravitational field of the Moon is the result of the distribution of mass within the Moon. Information on the gravity, combined with density and topographic data provides clues to internal properties and provides data for models of composition and degree of differentiation. Anomalies in the gravity field around the planet lead to interpretations of the nature of the lunar crust and mantle relationships.

Topographic profiles obtained from orbit during Apollo missions show that the highlands are rough compared with the maria. The centre of figure of the Moon is displaced by about 2·5 km from the centre of mass, in other words the farside is generally at higher elevations than the nearside. By contrast on a broad scale the lunar gravity field is rather smooth, except over the ringed maria. Kaula *et al.* (1974) argue that because of this difference between gravity and topography, isostasy prevails. They argue that there is a crust of lower density, say equivalent to anorthosite (2950 kg m^{-3}), of 45–60 km thickness, and that this is thicker on the farside than on the nearside.

The Lunar Orbiter missions, although primarily photographic, yielded important data on the gravity field. Scientists at the Jet Propulsion Laboratory, in analysing the Doppler radar tracking of the spacecraft, noted that the spacecraft accelerated over certain regions of the Moon. The accelerations were attributed to differences in concentrations of mass within the Moon, and these anomalies were therefore termed *mascons* (Muller and Sjogren, 1968). Since the discovery of mascons, all lunar spacecraft tracking data have been analysed to refine the gravity data. Because the method of mapping the gravity field required accurate tracking of the spacecraft from Earth (line-of-sight), detailed gravity data are available only for the lunar nearside.

Examination of the distribution of the mascons shows that positive anomalies, that is higher than average concentrations of mass, are associated with the circular basins of Imbrium, Serenitatis, Crisium, Smythii, Humorum, and Nectaris, and a weak positive anomaly for the centre of the Orientale basin. Mascons are also associated with Sinus Aestuum and Sinus Medii. Negative anomalies (i.e. areas with less than average mass distributions), occur in Mare Tranquillitatis, Mare Fecunditatis, Mare Nubium, Mare Vaporum, and in the Orientale basin in a lowland ring between the Rook Mountains and the Cordillera Mountains.

The association of positive gravity anomalies with the large impact basins provides a clue to the origin of mascons. The circular basins were formed before the emplacement of mare basalt, possibly in a volcanic crust of KREEP basalts (table 9.1) and anorthositic materials of lower density ($\rho = 2900 \text{ kg m}^{-3}$) than the sub-surface mantle ($\rho = 3300 \text{ kg m}^{-3}$). Isostatic rebound following the formation of a basin could have resulted in a plug of dense mantle material rising above the normal crust–mantle interface. At this stage, the basins (~ 8 km deep) would still not show a gravity anomaly. However, the dense mare basalts (3300–3400 kg m^{-3}), emplaced in the basins about 3700 million years ago, provide an 'excess mass' on top of the plugs to form mascons. The mass distribution in the Orientale basin suggests that flow might have been lateral from the outer ring toward the centre, leaving an outer negative ring. This explanation for mascons is one of several that have been put forward.

Gravity data on a scale smaller than the circular basins show that craters smaller than about 100 km in diameter have negative anomalies. Lunar gravity data on a local scale were obtained during the Apollo 17 mission by the Traverse Gravimeter Experiment in the Taurus-Littrow valley. Preliminary results from this experiment indicate that the valley is floored with a basalt block about 1 km thick and that the valley sides are steep beneath the basalt.

11.4. *Lunar seismology*

Our understanding of the internal characteristics of Earth and its subdivision into core, mantle and crust is derived to a large extent from seismological data gathered from hundreds of stations around the globe. The establishment of seismic stations on the Moon has also played the dominant role in learning about the interior of the Moon (fig. 11.2). Lunar seismic data have been derived from three types of Apollo experiment:

(1) Passive Seismic Experiment, designed to monitor natural seismic events, deployed at Apollo sites 11, 12, 14, 15 and 16 (all but that at the Apollo 11 site are still working).

(2) Active Seismic Experiment, using artificially generated seismic devices (e.g. explosives) to determine shallow lunar crustal structure to a depth of several kilometres, deployed on Apollos 14 and 16.

(3) Lunar Surface Profiling Experiment, similar to the active seismic experiment, consisted of an array of geophones, electronics and eight explosive packages; deployed in the Littrow Valley to determine shallow crustal structure and meteoritic flux.

Lunar seismic signals are quite different from those of Earth and are of exceedingly long duration, showing gradual increases and decreases in intensity (fig. 11.1). Their characteristics seem to be explained best by the presence of a dry heterogeneous surface layer (probably regolith and fractured bedrock) in which seismic waves are highly scattered, underlain by a more Earth-like elastic non-scattering material. Because of the long duration of the signal ('ringing' effect caused by continued vibration of individual blocks in a fractured crust), later reflection and arrivals of waves other than the compressional wave are masked. Consequently, most of the seismic data for the Moon must be derived from compressional (primary) waves.

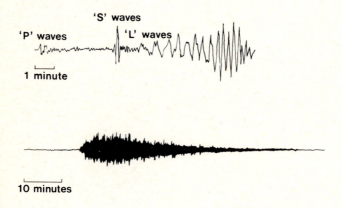

Fig. 11.1.   A typical lunar seismograph trace of a moonquake (below) compared with one from Earth. Note differences in scale.

Natural seismic events on the Moon appear to originate from four main sources: deep moonquakes, shallow moonquakes, high frequency teleseisms or 'quakes originating at long distances from the seismometer, and meteoritic impacts. The rate of release of seismic energy from moonquakes (2 MJ per year) is considerably less than from Earth (10^{11}–10^{12} MJ per year) both in numbers of events and their size, registering with maximum Richter magnitudes between 2 and 3.

Deep moonquakes originate at depths between 600 and 950 km near the lithosphere/asthenosphere boundary. The epicentres of the 37 foci so far located define two broad belts on the nearside (fig. 11.3). There is not obvious correlation of the epicentres with any surface features. The periodicity of moonquakes, however, correlates with the Moon's tidal cycle, suggesting that they are triggered by tidal forces. In contrast, shallow moonquakes correlate very well with sunset and sunrise and we consider them to be generated by thermal expansion and contraction of surface and near-surface rock.

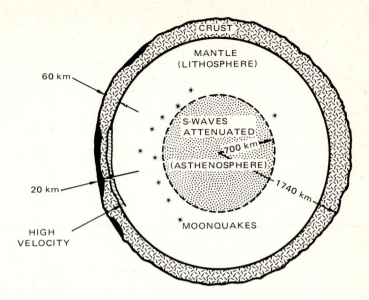

Fig. 11.2. Schematic diagram of the lunar interior showing zone of moonquakes, relation to marial basalts to the crust (crustal thickness is exaggerated) and possible limited extent of the seismic high velocity zone (cross-hatching). Lunar nearside is to the left of the figure (from Dainty *et al.*, 1973).

Although High-Frequency Teleseisms account for less than one per cent of all natural lunar seismic events, they are the largest and have signals similar to earthquakes. Unfortunately their foci all lie outside the seismic network and little is known about them except that they appear to be shallow, perhaps 300 km deep and less, and may be of tectonic origin.

The passive seismic network has detected an average of 75 to 100 events per year that appear to be meteoritic impact. The network is sufficiently sensitive to detect the impact of bodies as small as about 10 g near the network, or 1000 kg at the maximum distance from the net. The flux of meteorites derived from these records is 1 to 3 orders of magnitude less than that derived from Earth-based measurements; however, the discrepancy may result from poor calibration.

Analyses of signals from natural and man-made seismic events (explosives and impacts of used spacecraft) show that the interior of the Moon is multi-layered. The outer kilometre (fig. 11.4) has seismic compressional wave velocities that increase rapidly from about 100 m s^{-1} to about 900 m s^{-1} indicating the change from loose porous regolith and more consolidated material. There is then a marked

Fig. 11.3. Distribution of deep seismic epicentres (crosses) (after Latham *et al.*, 1973). Centres of main mascons marked with large dots.

increase in velocities from 900 m s^{-1} to 6 km s^{-1} to a depth of about 20 km, representing fractured basalt with the gradual increase being attributed to increased consolidation and crack closure.

At 20 km depth, the velocity increases abruptly to 6·7 km s^{-1} and remains fairly constant to 60 km depth, indicating that most of the cracks are 'healed' and that the temperature and pressure gradients are small, or that they cancel one another out. This velocity is consistent with an anorthositic gabbro composition.

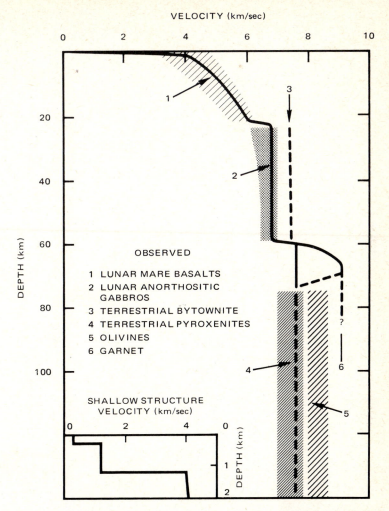

Fig. 11.4. Comparison between the compressional velocity profile for the lunar crust and upper mantle derived from seismic measurements, and the velocities of several types of lunar and terrestrial rocks. The velocity model is shown by a heavy line (or dashed heavy line where the model is uncertain) (after Dainty *et al.*, 1973).

At 60 km the velocity makes a marked increase to about $9 \, \text{km s}^{-1}$. The nature of this increase is poorly known and it may even be spurious; there are only a few exotic minerals in which such a high velocity could occur. This 60 km discontinuity is considered to be the boundary between the lunar crust and mantle; however, whether this crust is regional or Moon-wide has not been determined.

The apparent seismic velocity to a depth of at least 150 km is about 7.8 km s^{-1}, which is intermediate between terrestrial crustal and mantle velocities and is appropriate for pyroxenites.

Information on lunar seismic characteristics below 150 km depth (deepest penetration of seismic energy from artificial impacts) is limited. However, it appears that the compressional seismic wave velocities are fairly constant at 8 km s^{-1} to a depth of nearly 1000 km.

Below about 1000 km shear waves are strongly attenuated, indicating the presence of a 'soft' lunar core. Although the change in signal could be due to the presence of water or some other volatile, many lunar scientists believe it is the result of a partially melted (a few per cent of melt) core of a radius 600 to 900 km.

11.5. Heat flow

The measurement of heat escaping from the Moon is fundamental in placing constraints on models (fig. 11.5) of the composition and structure of the lunar interior (Langseth *et al.*, 1972). Two possible

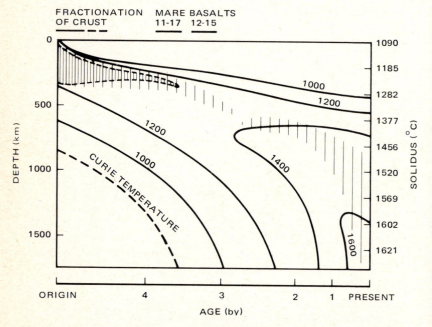

Fig. 11.5.   Thermal evolution of the Moon. Solid curves are isotherms in °C. Heavy shaded areas indicate greater than 50 per cent melt; light shaded areas less than 50 per cent melt. The Curie temperature of iron is shown by the dashed curve. Solidus temperatures are those used to define melt zones (from Hubbard and Minear, 1975).

sources of heat are production from radioactive decay and residual heat from the time of formation of the Moon. For a body as small as the Moon, even if it was initially molten, the present heat flux would be extremely small. Thus, most of the heat presently measured must presumably come from the disintegration of uranium, potassium and thorium.

The earliest measurements of lunar heat flow were made by Baldwin, who detected radiation from the Moon in the microwave band with Earth-based instruments. The first measurements *in situ* were made during Apollo 15 when a probe was placed in a borehole. Similar measurements were planned during Apollo 16, but the instrument was accidentally broken before data could be acquired. The second heat flow measurements were made on Apollo 17.

The heat flow measured at the Hadley Rille (Apollo 15) and Taurus–Littrow (Apollo 17) is about half the average heat flow of the Earth. If these measurements are typical of the Moon, then heat production *per unit mass* for the lunar interior is more than twice that of the Earth.

Both of the two heat flow measurements *in situ* were made on mare–highland boundaries and might be influenced by lateral compositional and physical changes. Thus, caution must be exercised in applying the results from these two sites to the Moon as a whole.

11.6. *Tectonic features*

Tectonic features are surface expressions of deformation of the Moon's surface layers. Based on spacecraft pictures, there are several types of feature that may be unambiguously related to tectonic activity. These are: the straight and arcuate rilles (fig. 11.6), consisting of down-faulted troughs bounded by straight or arcuate faults, mare ridges, and linear scarps. Based on telescopic pictures of the Moon, and before the impact mechanism achieved wide acceptance, many other features of the lunar surface were attributed to tectonic activity. Now explained as artifacts of impact, such features include secondary impact crater-chains and the lineated terrain surrounding impact basins.

Here we shall restrict our discussion to features clearly identified as expressions of tectonic activity. The most obvious of these are the rilles, each of which may extend for several hundred kilometres with average widths of about 1 km, the larger ones being five kilometres or more across. Generally the larger rilles are about 1 or 2 km deep. These features are interpreted as graben. The widths of the graben change as they run across irregular topography and become wider with greater elevation (fig. 11.7). This implies that the bounding faults are inclined inwards and careful measurements by McGill (1971) show that the

Fig. 11.6.   Arcuate rilles (graben) surrounding the east side of Mare Humorum (north at bottom). The two 'blobs' near the middle of picture are spacecraft processing errors. (NASA Lunar Orbiter IV 132 M.)

Fig. 11.7.   Oblique photographs of the Ariadaeus rille taken by the Apollo 10 astronauts. This rille is a graben and shows the widening of the graben over a topographic high, indicating that the bounding faults are inclined inwards. (NASA Apollo 10–31–4646.)

inclination of the faults bounding the graben is about 65°, comparable to that of graben-bounding faults on Earth. Many graben are seen to cut mare materials but are overlain by younger mare units or post-mare craters suggesting that the graben were formed during and just after the time of mare emplacement. They often form well defined patterns on

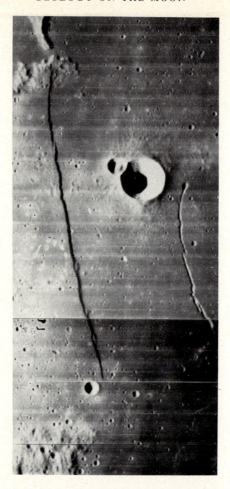

Fig. 11.8.   The Straight Wall—a lunar normal fault about 100 km long. North at bottom.
(NASA Lunar Orbiter IV 113 H_1 and H_2.)

the lunar surface; for example, the arcuate fractures are normally concentric to the major basins. Straight rilles such as the Ariadaeus and Hyginus rilles (fig. 7.7) are associated with collapse pits possibly indicating their association with volcanic activity.

While the rilles clearly indicate brittle fracture of the lunar surface, the mare wrinkle ridges are at least in part a result of buckling of the surface layer, accompanied by some fracturing. The origin of mare ridges has been discussed in more detail (Chapter 4) and it is possible that they are due, at least in part, to local movements of mare crust over sheets of sub-surface liquid magma. However, correspondence between mare ridge patterns and graben may suggest that tectonic activity also played a part in buckling the mare surfaces.

Single normal faults are rare on the Moon but one striking example is the Straight Wall (fig. 11.8) in Mare Nubium. According to Fielder (1965) the throw on this fault is 340 ± 15 metres; the angle of slope of the scarp is about $10°$, but this slope almost certainly represents a talus slope rather than the dip of the fault plane. Major thrust faults are not seen (cf. Mercury, Chapter 12).

Some of the individual tectonic features or complexes can be interpreted in terms of local movements. For example the concentric patterns around the margins of the maria in the circular basins probably result from down-sagging of the marial plate towards the centre of the basin. However, several attempts have been made to find Moon-wide patterns explicable in terms of global tectonics. Most of these were based on plotting various linear features, many of which would not be considered as tectonic today. Such attempts at a global synthesis resulted in the recognition of the *Lunar Grid Pattern*, so named by Spurr because of the tendency for lunar lineaments to form an orthogonal pattern of features running north-east and north-west. Other subsidiary patterns such as a north–south set of lineaments and radial and concentric patterns close to the circular basin were also identified. Studies by Fielder (1963) and Strom (1964) illustrate these patterns. However, all these studies were made before the advent of space photography and are thus restricted to the nearside of the Moon. Fielder considered that most of the features identified represent strike-slip, faulting, although spacecraft pictures have provided little evidence for this, and both Fielder and Strom considered the pattern in terms of north–south compressional movements. Various explanations of a north–south compression have been put forward including gravitational interaction between the Earth and the Moon and the presence of a simple convection cell system (Fielder, 1965).

For a comprehensive understanding of the causes of the deformation of the lunar crust, we must probably look for a less simple explanation consisting of combinations of more than one process operating at different times in lunar history. In an attempt to unravel this story, Mason *et al.* (1976) mapped all the graben-type rilles and obvious faults (fig. 11.9). These were chosen for study because they are unambiguously of tectonic origin. It became clear that many of these features were concentric or radial to circular impact basins although in most cases such patterns were restricted to the vicinity of the basin. However, a plot of these features on a stereographic projection centred on the largest of the lunar basins, Imbrium, shows that at least 30 per cent of

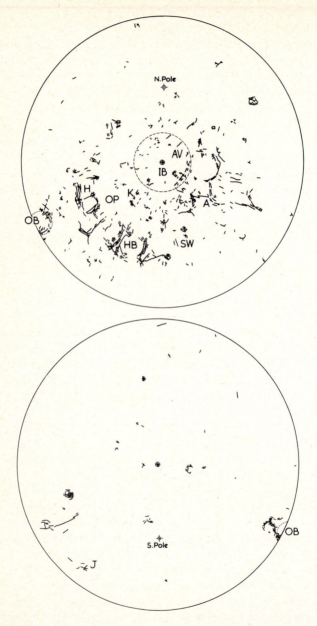

Fig. 11.9. Distribution of graben plotted on a stereographic projection centred on
Imbrium. IB, Imbrium basin; AV, Alpine valley; OP, Oceanus Procellarum;
OB, Orientale basin; HB, Humorum basin; SW, Straight Wall; H, Hevelius; K,
Kepler; J. Janssen. (From Mason *et al.*, 1976.)

the total length of graben are either concentric or radial to the Imbrium Basin. This was not a local pattern but extended over more than one hemisphere of the Moon. Such a pattern is to be expected from a major impact event. However, the graben that we see now at the surface cut mare materials that are considerably younger than the Imbrium event and Mason *et al.* suggest that the Imbrium event imposed a joint pattern in the lunar lithosphere and that later movements caused by Earth–Moon tidal interaction, volcanism and major impacts reactivated these fractures to give the present Imbrium pattern. It was also noted that many of the graben, even those associated with Imbrium, correspond roughly to an orthogonal grid pattern.

On the basis of our understanding of the Moon, therefore, there are three processes that may have contributed to the present lunar tectonic pattern. These are:

(1) Fracture patterns associated with the impacts forming the larger lunar basins.

(2) Movements generated by processes associated with igneous activity including tensional movements in the near-surface crust, deep-seated movements produced by magma generation and possibly convection.

(3) Tidal movements related to gravitational interaction between the Earth and Moon, possibly at a time when the Moon was nearer the Earth. These are still active as evidenced by seismic studies.

Each of these processes may have operated at different times in lunar history and any tectonic synthesis should take them all into account.

11.7. *Summary*

Throughout this chapter the statement has been made that various geophysical measurements place constraints on models of the lunar interior, structure and composition. Hubbard and Minear (1975) have drawn these data together to derive a general model of the thermal evolution of the Moon, portrayed in fig. 11.5. In their model, the outer 300–400 km of the Moon was initially molten and differentiated to form a plagioclase-rich crust containing a high percentage of uranium. By the time the marial basalts were extended, this crust was not only rigid enough to support the denser basaltic mascons, but had also developed some tensional faults, evidenced by some of the linear rilles.

With time, radioactive heating progressively warmed deeper parts of the lunar interior, ultimately producing a partly molten 'asthenosphere' below 800 km at present but possibly within 100–200 km from the surface 3000 million years ago. Deep moonquakes appear to be associated with the lithosphere/asthenosphere boundary.

12. Beyond the moon

12.1. *Introduction*

The Moon has served as an excellent training ground to learn techniques for extraterrestrial exploration. However, the field area of the planetary geologist is much wider than just the Moon—he has the entire solar system as his bailiwick, or at least the solid surface planets and satellites (Table 12.1).

During the 1960s and early 1970s planetary geologists were split generally into two groups; those who looked primarily at the Moon, and those who looked at the rest of the planets. With the tremendous advances in space flight and the high frequency of missions during the period, lunar and planetary missions were run concurrently and it was impossible for any single geologist to be involved intensely with more than one planet at a time.

With the ending of the Apollo series, and with planetary missions less frequent, there is now time to analyse in depth the tremendous wealth of data that have been collected. Now is the time for comparing the geology of one planet with another and to view the terrestrial planets, including the Moon, as covering a wide spectrum of geological characteristics and histories. We are speaking here of comparative planetology in which we analyse and interpret differences and similarities among the planets in terms of geological processes and histories.

Although most geological data from the inner planets have come from U.S. missions, principally Mariner spacecraft, some important information has been obtained by Soviet Missions and from Earth-based instruments.

12.2. *Mars—the red planet*

After Earth and Moon, the planet that we know most about is Mars. An excellent pre-Viking summary of the geology of Mars is given by Mutch *et al.* (1976).

Although Mars has been observed telescopically for over 350 years little of the information so gained could be used to develop a geological understanding of the planet. From telescopic observations it was known

Table 12.1. Data for comparison of planets in the Solar System.

	Mercury	Venus	Earth	Mars	Jupiter	Saturn	Uranus	Neptune	Pluto
Mean distance from Sun (millions of kilometres)	57·9	108·2	149·6	227·9	778·3	1427	2869·6	4496·6	5900
Period of revolution	88 days	224·7 days	365·26 days	687 days	11·86 years	29·46 years	84·01 years	164·8 years	247·7 years
Inclination of axis	<28	3	23° 27'	23° 59'	3° 05'	26° 44'	82° 5'	28° 48'	?
Equatorial diameter (kilometres)	4880	12 104	12 756	6787	142 800	120 000	51 800	49 500	6000 (?)
Mass (Earth = 1)	0·055	0·815	1	0·108	317·9	95·2	14·6	17·2	0·1 (?)
Volume (Earth = 1)	0·06	0·88	1	0·15	1316	755	67	57	0·1 (?)
Relative density	5·4	5·2	5·5i	3·9	1·3	0·7	1·2	1·7	?
Atmosphere (main components)	None	Carbon dioxide	Nitrogen, oxygen	Carbon dioxide, argon (?)	Hydrogen helium	Hydrogen, helium	Hydrogen, helium, methane	Hydrogen, helium, methane	None detected
Mean temperature at visible surface (degrees celsius) S = Solid, C = Clouds	350(S) day −170 (S) night	−33 (C) 480 (S)	22 (S)	−23 (S)	−150 (C)	−180 (C)	−210 (C)	−220 (C)	−230 (?)
Atmosphere pressure at surface (millibars)	10·9	90 000	1000	6	?	?	?	?	?
Surface gravity (Earth = 1)	0·37	0·88	1	0·38	2·64	1·15	1·17	1·18	?
Known satellites	0	0	1	2	13	10	5	2	0

Fig. 12.1.   Viking Orbiter 1 photomosaic of Mars showing the southern hemisphere.
The Argyre impact basin is surrounded by a ring of mountains. Cratered terrain of
the type photographed by Mariners 6 and 7 is seen in the foreground. (NASA
Viking photograph.)

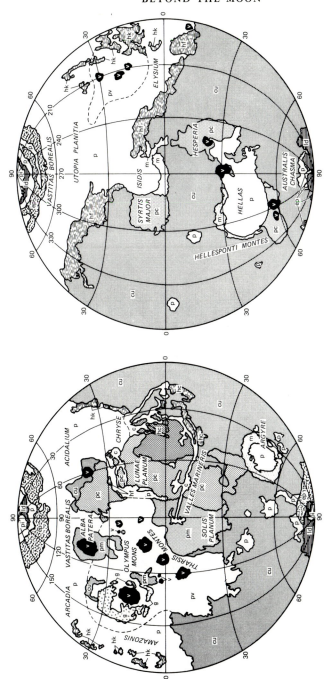

Fig. 12.2.   A simplified physiographic map of Mars. (After Mutch *et al.*, 1976.)

that Mars has an atmosphere, that white cloud patterns developed, and that dust-storms were a frequent occurrence, the larger ones enveloping almost the whole planet. It was also known that Mars had ice caps which retreated and advanced with the seasons.

The Red Planet was finally brought into the realm of the geologist in 1965 with the successful fly-by of Mariner 4, its single television camera returning a total of 22 pictures during its brief 30 minute encounter with Mars. The pictures were rather poor quality and had at best a ground resolution of about 3 km; nonetheless, it was the first close-up view and the pictures revealed a cratered terrain not unlike that of the Moon. This led many scientists to view Mars as being simply another relatively inactive planet like the Moon. The successful fly-bys of Mariner 6 and 7 in 1969 appeared to confirm this view (fig. 12.1) with a better than 10-fold increase in surface resolution and coverage. However, some large scale pictures did show several types of landform that do not exist on the Moon and which gave cause to the question just how Moon-like Mars actually is. These landforms included irregular etched surfaces, chaotic terrain consisting of huge masses of jumbled blocks, and craters with morphologies clearly different from lunar craters. All these landforms suggested that although Mars was a Moon-like body, largely covered with craters, some internal processes might have operated to modify the surface. Based on these observations the view of Mars in 1970 was of a planet with an ancient crust battered in its early history by meteoritic impacts with some minor modifications suggesting that the interior had been active in more recent times.

Mariner 9, put in near-polar orbit in 1971 to investigate all of the martian surface, demonstrated the danger of drawing planet-wide conclusions from investigations of only a limited part of the surface. When Mariner 9 first arrived at Mars there was little to see because the surface was shrouded by a tremendous dust storm. As the storm abated the upper atmosphere began to clear and topographically high features such as the large volcanoes began to appear. This was the first evidence that Mars was not the planet expected from previous missions. These volcanoes were the first of many surprises to come throughout the mission as the entire surface became available for observation.

All the pre-Mariner 9 pictures had been taken of the southern hemisphere which indeed is old and densely cratered, but pictures by Mariner 9 of the northern hemisphere showed a completely different terrain (fig. 12.2). In this region most of the old cratered surface has been buried, presumably by volcanic and wind-blown material, and there are large volcanoes rising high above the surface (fig. 12.3). High resolution pictures showed dramatic evidence of internal activity,

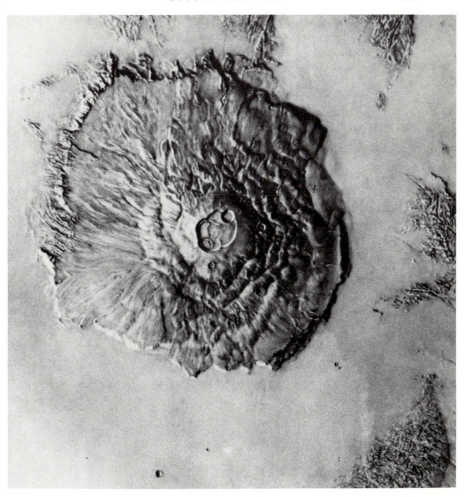

Fig. 12.3. An airbrush rendition of Olympus Mons, a large martian volcano with a base diameter of 600 km and a height of about 26 km above the surrounding terrain. Note the scarp around much of the outer margin. (Courtesy M. Carr and U.S. Geological Survey.)

there being complicated fracture patterns and features such as the Valles Marineris rift, hundreds of kilometres long and tens of kilometres across (fig. 12.4). In addition, there is abundant evidence of external modification by wind erosion and possibly fluvial activity. Mariner 9 therefore dramatically changed the 'dead planet' view of Mars to one in which there is a long surface history of internal and external modification possibly right up to the present time.

The oldest recognizable structures on Mars are the circular basins (fig. 12.1) some of which are multi-ringed. In many respects these are the martian counterparts of lunar circular basins with which they have many features in common. The largest basin is Hellas which is slightly elliptical having a maximum diameter of about 2200 km and a depth of 4 km. Apparently this basin acts as a huge dust bowl and was one of the last regions to clear during the 1971 dust storm.

As on the lunar highlands the large basins are associated with cratered terrain which occupies much of the southern hemisphere. Craters in this terrain tend to be less fresh than those on the Moon and there is little evidence of ejecta or secondary cratering activity, suggesting that erosion or modification has been much stronger on Mars than on the Moon, as expected in the presence of an atmosphere, albeit tenuous.

Once the early phase of bombardment ended, the martian crust continued to develop. The most obvious evidence of internal activity is the presence of large young shield volcanoes (Carr, 1976), the most striking of which is Olympus Mons (fig. 12.3). This gigantic volcano is more than 600 km across and rises to 6 km above the surrounding plane. Although the general morphology is similar to that of Mauna Loa (Hawaii), the largest volcano on Earth, it is several times larger. The summit of Olympus Mons is marked by an impressive caldera complex more than 80 km across. The flanks of the shield consist of thin narrow flows, flow lobes, lava channels and partly collapsed lava tubes similar to those seen on basaltic terrains on Earth. One of the enigmatic features of Olympus Mons is a prominent scarp, up to 2 km high, that surrounds much of its base. The origin of this scarp may be erosional or tectonic. Other volcanoes occur in the same region as Olympus Mons, the largest of which lie on the Tharsis ridge, a thickening of the martian crust more than 4000 km long. These volcanoes are slightly smaller and in some cases appear to be more degraded than Olympus Mons. Dome-like volcanoes occur at the northern end of the Tharsis ridge; these are smaller than the shield volcanoes and may result from different compositions of lava or different mechanisms of eruption.

Other evidence of volcanic activity includes numerous small conical hills with central summit craters, possibly pyroclastic cones. Extensive smooth plains may be vast deposits of volcanic ash and there is evidence of flood-basalt type volcanism analagous to lunar maria.

Tensional faulting is common in many parts of Mars. The Valles Marineris (fig. 12.4) appears to be a gigantic rift system more than 4000 km long. It is tempting to think that it represents the beginning of

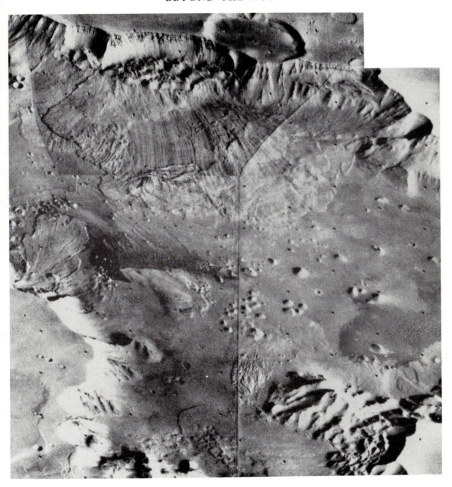

Fig. 12.4. Viking Orbiter oblique photomosaic across the martian canyonland. The
canyon shown here is about 2 km deep. Large landslides are seen on the far
canyon wall and a dark dune field to the lower right. Sunlight from right. (NASA
Viking photograph.)

plate separation which was not sustained. On a global scale, tectonic
activity on Mars may be summarized as follows:

(1) the uplift by many kilometres of the Syria rise (which includes the
    Tharsis ridge),
(2) the development of the Valles Marineris rift system,
(3) down-sagging of much of the northern hemisphere, now covered
    with younger deposits that bury the older cratered terrain.

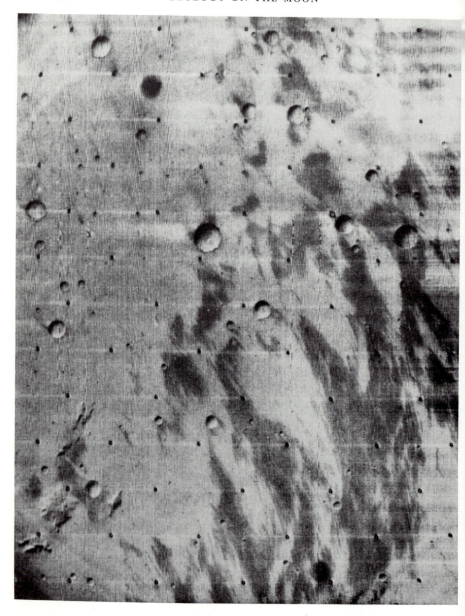

Fig. 12.5.   Albedo-streaks on Mars caused by action of the wind. The geometry of the craters causes differential wind erosion and deposition. (NASA Mariner 9 photograph.)

Fig. 12.6. Evidence of water erosion on Mars. The large channel running across the picture appears to have been eroded by water that originated in the chaotic collapsed area to the right. The distance across the picture is about 130 km. (NASA Viking Orbiter 1 photograph.)

The exciting picture that is emerging from Mars is one of a planet with a very complex and interesting history. The variety and scale of landforms show that it has been subject to many diverse geological processes including volcanism, tectonism and impact cratering, together with wind erosion and deposition (fig. 12.5) and possible fluvial activity (fig. 12.6). Unlike the Moon there may be evidence of incipient plate-type tectonism and large-scale uplift analogous to Earth-like mountain building.

12.3. Mercury

Nearly everything that is known about the geology of Mercury has been gained from a single spacecraft launched in 1973 as a fly-by mission —Mariner 10. Until this mission, Mercury was little more than a fuzzy disc seen through a telescope. Its optical properties were apparently like those of the Moon suggesting the presence of a regolith, and therefore that it, too, had been churned by impacts.

Mariner 10 arrived at Mercury in March of 1974 and as it sped past the planet at more than 10 km s^{-1} its twin television cameras began to acquire images of the surface—more than 2000 frames were transmitted during the first pass of the planet (Murray et al., 1974a, b). The trajectory of the spacecraft was such that after swinging by the Sun, Mariner 10 returned to Mercury a second time in September 1974, and then again in March 1975 greatly increasing the number of pictures. Nearly 50 per cent of the planet was imaged by Mariner 10 (fig. 12.7) at resolutions comparable to Earth-based views of the Moon.

Each terrestrial planet has certain characteristics in common with other planets, but at the same time appears to have had its own unique history. For example, Mercury has a density similar to that of the Earth, indicating a large iron core, but in other respects is quite unlike the Earth. Mariner 10 revealed a surface that is superficially Moon-like with heavily cratered terrain and regions of smooth plains. Closer examination however, shows that many important differences exist between the Moon and Mercury. The craters and basins have the same characteristics as lunar craters, some being fresh with extensive rayed systems and well defined ejecta sheets and fields of secondary craters, while older craters have lost these features to a greater or lesser degree depending on their age. Mercury, however, has a higher gravitational field than the Moon and thus the ballistic range of ejecta is less, causing secondary crater fields to be closer to their primary craters than on the Moon (Gault et al., 1975). The craters are also slightly shallower; central peaks and inner ring mountains appear on smaller craters than on the Moon—again effects that can be explained by the gravity difference.

Other features that appear to be peculiar to Mercury are sinuous scarps which have been interpreted as reverse faults (Strom *et al.*, 1975). During its early history the radius of Mercury may have become smaller by a few kilometres, causing the crust to buckle in the form of reverse faults.

The major geological provinces for Mercury (fig. 12.8) have been mapped by Trask and Guest (1975). The oldest units are heavily cratered terrain and inter-crater plains, although the relationship between these two terrains is not clear. One interpretation is that the inter-crater plains represent a primordial planetary surface that was cratered during a phase of intense bombardment; because the ejecta from individual craters is far less extensive than for similar craters on the Moon, not all the primordial crust was affected. In many areas inter-crater plains have been modified by swarms of secondary crater chains derived from the large craters, but in other areas some inter-crater plains may be younger than the heavily cratered terrain and the distinction between different ages of inter-crater plain awaits more detailed study.

Superimposed on this old degraded terrain are younger basins and craters that are degraded but which still have recognizable ejecta and secondary ejecta fields. The youngest of the known major impacts is the 1300 km Caloris Basin (fig. 12.9), the most imposing structure seen by Mariner 10 on Mercury. In size and appearance it closely resembles the larger lunar basins and is similarly bounded by a discontinuous ring of mountains that stand on average about 2 km above the basin floor. A weak outer scarp occurs in places at about 150 km beyond the main scarp. A well-defined radial system of hills and grooves extends outwards from the basin's rim, and a hummocky ejecta unit composed of small rounded hills resembles deposits found around the lunar Orientale Basin.

Post-dating the formation of Caloris are extensive plains-forming units close to the basin and in the north polar region. Similar units occur on crater floors elsewhere on the planet. The origin of these smooth plains is an interesting and important problem. In some respects they are similar to the lunar Cayley plains and might consist of impact-generated debris that has been modified and smoothed by mass-wasting and seismic events. It is also suggested that they are analogous to the lunar maria and represent a period of extensive volcanism. Strom *et al.* (1975) argued that the presence of these plains units far from Caloris favours their being volcanic plains rather than

Fig. 12.7(*a*).   A photomosaic of the side of Mercury seen as Mariner 10 left the planet after the first encounter. On this face of Mercury the Caloris Basin is seen near the terminator (see fig. 12.9); areas of smooth plains surround Caloris and occur near the north pole (see fig. 12.8 for map). (NASA photograph.)

Fig. 12.7(*b*).   A photomosaic of the face of Mercury seen as Mariner 10 approached the planet during the first encounter. Much of this face of the planet is cratered and resembles the lunar highlands. (NASA photograph.)

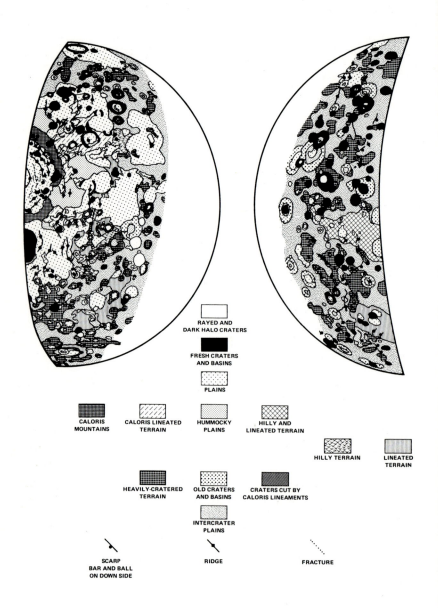

Fig. 12.8.   Geological/terrain map of Mercury (from Trask and Guest, 1974).

Fig. 12.9. A photomosaic of the 1300 km-diameter Caloris Basin on Mercury. This
   picture includes the highest resolution pictures from all three Mariner 10
   encounters from Mercury. (NASA photograph.)

impact melts. Clearly the presence or absence of volcanism is an
important question since it bears on the thermal history of the planet.

The later history of Mercury, like that of the Moon, was relatively uneventful apart from occasional impacts producing large craters.

We must bear in mind that only half of Mercury has been observed, and based on our experience from Mars we must be cautious in generalizing for the whole planet from the present data. Even for the half of the planet we have seen (fig. 12.8) the distribution of terrain types is assymetrical.

12.4. *Venus*

Despite the proximity of Venus to the Earth, less is known about it than any of the terrestrial planets, particularly with regard to its surface structure. Yet the diameter and mass of Venus are nearly the same as the Earth and in many respects it may be our closest analogue. Typically, the exploration of a planet begins with reconnaissance missions that acquire images of its surface. Because the surface of Venus is completely masked from conventional optical imaging systems by its ubiquitous cloud cover, such a reconnaissance has not been feasible.

Significant data for the complex venusian atmosphere have been collected from several Soviet probes and U.S. fly-bys. Most of our geological data for Venus have come from Soviet Venera spacecraft (landers) and Earth-based radar observations.

The atmosphere of Venus consists primarily of carbon dioxide and is extremely dense, its surface pressure being nearly 100 times greater than that of Earth. The same cloud cover that masks the surface from view also appears to trap solar radiation and results in extremely high surface temperatures which average more than 500 °C—so hot that lead would be molten! Thus, high temperature and pressure make the surface of Venus rather hostile and a nasty place to land spacecraft. They do, however, offer the potential for some interesting and bizarre processes which scientists have only begun to think about.

The only way to see the surface of Venus is by radar with its long wavelength radiation capable of penetrating the clouds. Radar pictures have reached a high level of quality—most geologists are familiar with SLAR (Side Looking Airborne Radar) images that have been available for Earth since the mid-1960's. SLAR images have been used successfully in several major mapping programmes on Earth—most notably in the Amazon Basin where adequate maps had never been available because of dense cloud cover.

Although they are not exactly the same as SLAR images, radar maps for Venus have been produced by the giant radar facilities of Arecibo,

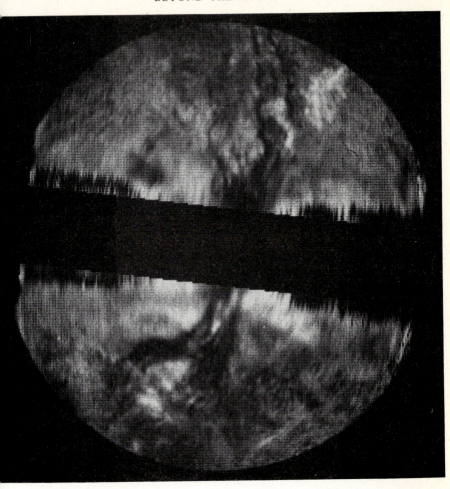

Fig. 12.10. Radar image of the surface of Venus acquired at the Goldstone Tracking Station. The image shows an area approximately 1500 km in diameter. A large, trough-like feature ($\sim$1500 km long by 150 km wide by 2–4 km deep) is seen near the centre of the frame. The main linear branch of the trough extends some 650 km trending NNE–SSW and is centred near the equator and $-76°$ east longitude. North of this linear branch the single-channel form deteriorates into several north-trending, roughly parallel troughs, the largest of which has scalloped walls, suggesting a crater-chain-like appearance. To the south, the trough divides into two branches, each about 100 km wide. A ridge or septum appears to run the length of the linear portion of the trough system. Surrounding the troughs are essentially featureless plains. By analogy with the East African/Ethiopian rift system on Earth and the Valles Marineris canyon system on Mars, the venusian feature could be interpreted to be the result of extensional tectonics. The dark horizontal band is an artifact of the radar process. (Radar image courtesy of R. Goldstein, Jet Propulsion Laboratory, caption by M. Malin.)

Puerto Rico and Goldstone, California. Fig. 12.10 is a radar map of Venus acquired at Goldstone. Differences in the grey-scale indicate differences in radar reflectivity which result from a whole host of complex and often inter-related surface properties. Among these properties are local topography, surface roughness, particle size and chemical composition. These maps yield useful information about the character of the venusian surface, especially when they are combined with radar-derived topographic data. For example, a feature known as Alpha is a circular feature more than 2000 km in diameter, putting it in the same size class as the Hellas basin on Mars, the Caloris basin on Mercury and the larger circular basins of the Moon. If our past experience is any guide, then Alpha, too, probably represents an early-formed impact basin.

A great many other circular features several hundred or more kilometres across have been identified. Many of these are confirmed by radar topographic profiles to be depressions, and some, when viewed with higher resolution have distinctly raised rims. Thus, there is the possibility that regions similar to the heavily cratered units of other terrestrial planets also exist on Venus. Possible tectonic features are also known. A linear depression about 200 km wide and at least 1200 km long is reminiscent of the martian canyonlands. One region of Venus is a topographically high area more than 6000 km long and 500 km wide—comparable, perhaps to the martian Tharsis ridge. In yet another area, a narrow, linear scarp-like feature several hundred kilometres long has been identified.

The view that is beginning to emerge for Venus is a surface made up of diverse physiographic features, including canyons, circular basins, cratered terrain, scarps and mountains. This view has been made possible with radar images with, at best, 15 km resolution. Improvements in radar facilities promise to yield radar images of 4 to 6 km resolution for more than ten per cent of the surface of Venus by the early 1980's. This increased capability can serve as the reconnaissance phase in normal planetary exploration, in preparation for a radar mapping spacecraft mission.

The first successful landing on Venus was accomplished by the Soviet Venera 7 spacecraft in 1970. Venera 7 transmitted about 23 minutes of data which included measurements of the surface temperature and pressure before the spacecraft 'died'. Venera 7 was followed in July 1972 by Venera 8 which successfully landed and transmitted 50 minutes of data, including an analysis of the surface by a gamma ray spectrometer which showed up to 4 per cent potassium, 200 p.p.m. uranium and 650 p.p.m. thorium. This analysis suggests to many investigators a possible granitic composition and the implication of planetary differentiation.

In 1975, Veneras 9 and 10 successfully landed on Venus and, as well as making measurements like those previously made, also took pictures showing the Venusian surface to be blocky. Additional and more refined measurements *in situ* are obviously needed for Venus, and all such measurements must be placed in a regional context derived from geological mapping.

12.5. *Comparative planetology*

In general, each time a planetary object is explored, three basic questions are asked (Greeley and Carr, 1976):

(1) what is the present state of the planet?

(2) how does the present state differ from previous states?

(3) how do present and previous conditions differ from those of other planets?

We now have direct geological information for Earth, Moon, Mars and Mercury, plus limited information for Venus, but we are a long way from answering the questions posed above. Nonetheless, we are beginning to gain insight into geological questions on a solar system scale—or at least an inner solar system scale. Just as chemists found no 'new' elements in the lunar samples (none were expected, but it is a question commonly asked), neither has the geologist discovered any 'new' geological processes. Igneous activity, tectonism, erosion and impact cratering remain the 'Big Four'. Evidence for all four is observed on all solid-surface planets. The exciting part of planetary geology is to see how the relative importance of these basic processes differs from planet to planet in response to different environments, and to try to understand the consequences of those differences. The differences in crater morphology resulting from different gravities on Mercury and the Moon illustrate the point. Similarly, the presence of an atmosphere greatly determines the role that gradation will play in shaping a planetary surface.

It is not the purpose here to enumerate all the differences and similarities exhibited on the planets. Rather, in this chapter we have briefly summarized the present geological knowledge for the planets and made comparisons to the Moon where appropriate. With each space mission, the field of planetary geology becomes more challenging. Even while this book was being published, new pictures of Mars were coming back from Viking Orbiters and Landers (fig. 12.11) giving us a new view of this exciting planet.

Fig. 12.11. The first picture ever taken from the surface of Mars. The middle of the picture is about 1·4 m from the Viking Lander camera that took the picture. The rock in the middle is about 10 cm across. (NASA photograph.)

Epilogue

Although this book has been directed towards the geology of the Moon, it should have become clear from the last chapter that study of the Moon is only a stepping stone to the geology of the whole of the solar system. The Moon is relatively easy to interpret and has proved an excellent testing ground for geological techniques applied to planetary bodies. In extraterrestrial geology much emphasis has to be placed, from necessity, on remote sensing techniques, but, in addition, on the Moon there has been the advantage of surface exploration to provide what is commonly called 'ground truth', that is a verification of the interpretations based on remote sensing. The validity of various methods of investigation including not only photographic images but also radar, infrared, gamma-ray and X-ray fluorescence techniques has been checked. The relative values of the different methods have been determined, and the stage is now set for specific remote sensing tasks to be completed from future lunar orbiters.

Testing of techniques on the Moon has provided geologists with the necessary confidence to apply these techniques to other terrestrial planets. As demonstrated in the exploration of Mercury, the preliminary geological mapping based on Mariner 10 pictures in 1974 was completed relatively soon after the first fly-by. Without the accumulated experience of lunar work this task almost certainly would have taken much longer. Exploration of the planets has only just begun and although the ultimate aim of geologists will be to handle rock samples collected from all of the solid planets, it is now clear that much geology can be determined by remote sensing techniques before samples become available for study. We are thus in an era when geological principles determined on Earth and tested on the Moon will play an important part in understanding the historical development of the whole solar system.

An ultimate aim both of astronomy and geology is to understand the origin of the planets in the solar system; but the approach taken by these two fields of science is quite different. While the astronomer attempts to reconstruct the sequence of events that led to the present array of planets around the Sun, the geologist attempts to work backwards from the present to the time when planets were formed. The astronomer bases his work on the known laws of physics and chemistry, and by making assumptions of the initial starting conditions he attempts to trace the sequences of events that may have led to the formation of the stars and their planetary companions. Such a method of reasoning is inevitably a theoretical one. On the other hand, a geologist attempts to establish the nature of processes that have occurred within the planet and to trace the history of these processes back in time, using observations and interpretations of the rocks that make up the planets. Ideally these two approaches to the understanding of the origin of our solar system should meet and overlap at the time of planetary formation, but inevitably there is a great gulf of time between them. Our geological studies of the planets now take us much further back in time than was possible from studies of Earth—tantalizingly close to the time of planetary origin—but we still cannot go back as far as we would like. Astronomical studies can provide the mechanisms of star formation and the type of environment that planets might have been formed in, but still fall far short of defining precisely the way in which individual planets were formed. Perhaps we shall never completely close the gap but the opportunity to try is clearly with us and shows that geological principles may lead the modern geologist into fields undreamed of by our geological forefathers.

Further reading

MUTCH, T., 1972. *Geology of the Moon: A Stratigraphic View*. Princeton University Press, Revised Edition.
Although the original edition of this well written book was prepared before the Apollo landings, the revised edition has been updated to include Apollo material.

SCHULTZ, P. M., 1976. *Moon Morphology*. University of Texas Press.
A detailed and well illustrated account of lunar geomorphology based on Lunar Orbiter pictures.

TAYLOR, S. R., 1975. *Lunar Science: A Post Apollo View*. Pergamon.
The author has accomplished the mammoth task of reviewing the present state of knowledge from studies of lunar samples.

BOWKER, D. E. and HUGHES, J. K., 1971. *Lunar Orbiter Photographic Atlas of the Moon*. NASA SP–206.
A compilation of photographs covering much of the lunar surface.

SHORT, N. M., 1975. *Planetary Geology*. Prentice Hall Inc.
Includes much lunar geology as well as important work on other terrestrial planets.

HARTMANN, W., 1972. *Moons and Planets*. Boyden & Quigley, Inc.

MACDONALD, G. A., 1972. *Volcanoes*. Prentice Hall Inc.

References

BALDWIN, R. B., 1963. *The Measure of the Moon*, University of Chicago Press.
BOWELL, E. L. G., 1971 a. Astronomy of the Earth–Moon system. In *The Earth and its Satellite* (Ed. Guest, J. E.), Chapter 2. London: Rupert Hart-Davis.
BOWELL, E. L. G., 1971 b. Polarimetric studies. In *Geology and Physics of the Moon* (Ed. Fielder, G.), Chapter 9. Amsterdam: Elsevier.
CARR, M. H., 1974. The role of lava erosion in the formation of lunar rilles and martian channels. *Icarus*, **22**, 1–23.
CARR, M. M., 1976. The volcanoes of Mars. *Scientific American*, **234** (no. 1), pp. 33–44.
Code of Stratigraphic Nomenclature, 1970. American association of Petroleum Geologists.
DAINTY, A. M., TOKSÖZ, M. N., SOLOMON, S. C., ANDERSON, K. R. and GOINS, N. R., 1974. Constraints on lunar structure. *Proc. 5th lunar Sci. Conf., vol. 3*, pp. 3091–3114.
DÄNES, Z. F., 1972. Dynamics of lava flows. *J. Geophys. Res.*, **77**, 1431–8.
DENCE, M. R., 1968. Shock zoning at Canadian craters. Petrography and structural implications. In *Shock Metamorphism of Natural Materials* (Eds. French, B. M. and Short, N. M.), pp. 169–184. Baltimore: Mono Books.
DENCE, M. R., 1971. Impact melts. *J. Geophys. Res.*, **76**, 5552–5565.
DENCE, M. R., 1972. The nature and significance of terrestrial impact structures. *24th Int. Geol. Conf. Section 15*, pp. 77–89.
DENCE, M. R. and PLANT, A. G., 1972. Analysis of Fra Mauro samples and the origin of the Imbrium Basin. *Proc. Third Lunar Science Conf., Supplement 3. Geochim. Cosmochim. Acta*, **1**, 379–399.
DENIS, J. C., 1971. Ries Structure, Southern Germany, A Review. *J. Geophys. Res.*, **76**, 5394–5406.
EGGLETON, R. E. and SCHABER, G. G., 1972. Cayley Formation interpreted as basin ejecta. *Apollo 16 Preliminary Science Report*, NASA SP–315, Pt. 29–7.
FIELDER, G., 1963. Lunar tectonics. *Q. J. Geol. Soc. Lond.*, **119**, 65–94.
FIELDER, G., 1965. *Lunar Geology*. London: Lutterworth.
FRENCH, B. M., 1968 a. Shock metamorphism as a geological process. In *Shock Metamorphism of Natural Materials*. (Eds. French, B. M. and Short, N. M.), pp. 1–17. Baltimore: Mono Books.
FRENCH, B. M., 1968 b. Sudbury Structure, Ontario: some petrographic evidence for an origin by meteorite impact. In *Shock Metamorphism of Natural Materials*, pp. 383–412. Baltimore: Mono Books.
FRENCH, B. M., 1970. Possible relations between meteorite impact and igneous petrogenesis, as indicated by the Sudbury structure, Ontario, Canada. *Bull. Volcanol.*, **34**, 466–517.
GAULT, D. E., 1970. Saturation and equilibrium conditions for impact cratering on the lunar surface: Criteria and implications. *Radio Science*, **5**, 273–291.
GAULT, D. E., 1974. Impact cratering. In *A Primer in Lunar Geology* (Eds. Greeley, R. and Schultz, P.), NASA Ames Research Center. NASA TMX 62359, pp. 137–143.
GAULT, D. E., GUEST, J. E., MURRAY, J. B., DZURISIN, D. and MALIN, M., 1975. Some comparisons of impact craters on Mercury and the Moon. *J. Geophys. Res.*, **80**, 2444–2460.
GAULT, D. E., HÖRZ, F. and HARTUNG, J. B., 1972. Effects of micro-cratering on the

lunar surface, *Proc. 3rd Lunar Sci. Conf., Geochim. Cosmochim. Acta,* Supplement 3, vol. 3, pp. 2713–2734.

GAULT, D. E., HÖRTZ, F., BROWNLEE, D. E. and HARTUNG, J. B., 1974. Mixing of the lunar regolith. *Proc. 5th Lunar Sci. Conf.,* vol. 3, pp. 2365–2386.

GAULT, D. E., QUAIDE, W. L. and OBERBECK, V. R., 1968. Impact cratering mechanics and structures. In *Shock Metamorphism of Natural Materials* (eds. French, B. M. and Short, N. M.), pp. 87–89. Baltimore: Mono Book Corp.

GILBERT, G. K., 1893. The Moon's face: A Study of the origin of its features. *Phil. Soc. Wash. Bull.,* **12,** 241–292.

GREELEY, R., 1971. Observations of actively forming lava tubes and associated structures, Hawaii. *Mod. Geol.,* **2,** 207–223.

GREELEY, R., 1976. Modes of emplacement of basalt terrains and an analysis of mare volcanism in the Orientale Basin. *Proc. 7th Lunar Sci. Conf.*

GREELEY, R., and CARR, M. H. (Eds.), 1976. A geological basis for the exploration of the planets. *NASA Special Publication,* **417,** 110 p.

GREELEY, R. and GAULT, D. E., 1971. Endogenetic craters interpreted from crater counts on the inner wall of Copernicus. *Science,* **171,** 477–479.

GREELEY, R. and HYDE, J. H., 1972. Lava tubes of the Cave basalts, Mount St. Helens, Washington. *Geol. Soc. Amer. Bull.,* **83,** 2397–2418.

GREEN, J., 1971. Copernicus as a lunar caldera. *J. Geophys. Res.,* **76,** 5719–5731.

GUEST, J. E., 1973. Stratigraphy of ejecta from the lunar crater Aristarchus. *Geol. Soc. Amer. Bull.,* **84,** 2873–2894.

GUEST, J. E. and MURRAY, J. B., 1969. Nature and origin of Tsiolkovsky crater, lunar farside. *Planet Space Sci.,* **17,** 121–141.

GUEST, J. E. and MURRAY, J. B., 1971. A large scale surface pattern associated with the ejecta blanket and rays of Copernicus. *The Moon,* **3,** 326–336.

GUEST, J. E. and MURRAY, J. B., 1976. Volcanic features of the nearside equatorial lunar maria. *J. geol. Soc. Lond.,* **132,** 251–258.

HARTMANN, W. K., 1973. Ancient lunar mega-regolith and sub-surface structures. *Icarus,* **18,** 634–636.

HAWKINS, G. S., 1964. *Meteors, Comets and Meteorites.* McGraw-Hill.

HAYS, J. F. and WALKER, D., 1974. Lunar igneous rocks and the nature of the lunar interior. *Proc. Soviet–American Conference on the Cosmochemistry of the Moon and Planets.* No. 208. (Lunar Science Institute, Houston.)

HEAD, J. W., 1974. Orientale multiringed basin interior and implications for the petrogenesis of lunar highland samples. *Moon,* **11,** 327–356.

HEAD, J. W., 1976. Lunar volcanism in space and time. *Rev. Geophys. Space Phys.,* **14,** 265–300.

HODGES, C. A., 1973. Mare ridges and lava lakes. *Apollo 17. Preliminary Science Report,* NASA SP-330, pp. 31–12 to 31–21.

HOLCOMB, R., 1971. Terraced depressions in lunar maria. *J. Geophys. Res.,* **76,** 5703–5711.

HÖRZ, F., MORRISON, D. A., GAULT, D. E., OBERBECK, V. R., QUAIDE, W. L., VEDDER, J. F., BROWNLEE, D. E. and HARTUNG, J. B., 1975. The micrometeoroid complex and evolution of the lunar regolith. *Proc. Soviet–American Conference on the Geochemistry of Moon and Planets.* Houston: Lunar Science Institute.

HOWARD, K. A., 1972. Ejecta blankets of large craters exemplified by King Crater. *Preliminary Science Report, Apollo 16.* NASA SP-315, pp. 29–70 to 29–77.

HOWARD, K., 1975. Geologic map of the Crater Copernicus. *Geologic Atlas of the Moon,* USGS I-840.

HOWARD, K. A., HEAD, J. W. and SWANN, G. A., 1972. Geology of Hadley Rille. *Proc. 3rd Lunar Science Conf., Geochim. Cosmochim. Acta,* **36,** Suppl. 3, 1–14.

HOWARD, K. A. and MUEHLBERGER, W. R., 1973. Lunar thrust faults in the Taurus–Littrow region. *Apollo 17 Preliminary Science Report,* NASA SP-330, pp. 31–22 to 31–25.

HOWARD, K. A. and WILSHIRE, H. G., 1972. Flows of impact melt at lunar craters. *J. Res. U.S. Geol. Survey,* **3,** pp. 237–251.

HOWARD, K. A., WILHELMS, D. E. and SCOTT, D. M., 1974. Lunar basin formation and highland stratigraphy. *Rev. Geophys. Space Phys.*, **12**, pp. 309–327.

HUBBARD, N. J. and MINEAR, J. W., 1975. A physical and chemical model of early lunar history. *Proc. 6th Lunar Sci. Conf., Geochim. Cosmochim. Acta*, suppl. 6, pp. 1057–1085.

HULME, G., 1973. Turbulent lava flow and the formation of lunar sinuous rilles. *Mod. Geol.*, **4**, 107–117.

JAMES, O. B., 1974. Lunar highland breccias generated by major impacts. *Proc. Soviet–American Conference on the Cosmochemistry of the Moon and Planets*, No. 212. Houston: Lunar Science Institute.

KAULA, W. M., SCHUBERT, G., LINGENFELTER, R. E., SJOGREN, W. L., WOLLENHAUPT, W. R., 1974. Apollo laser altimetry and inferences as to the lunar structure. *Proc. 5th Lunar Science Conf., Geochim. Cosmochim. Acta*, suppl. 5, 3049–3058.

KUIPER, C. P., 1965. Surface structure of the Moon. In *The Nature of the Lunar Surface* (Eds. Hess, W. N., Menzel, D. M. and O'Keefe, J. A.), Chapter 4. Baltimore: John Hopkins Press.

LANGSETH, M. G., CLARK, S. P., CHUTE, J. L., KEIHM, S. J. and WECHSLER, A. E., 1972. Heat flow experiment. *Preliminary Science Report, Apollo 15*. NASA SP–289, pp. 11–1 to 11–23.

LATHAM, G. V., EWING, M., PRESS, F., DORMAN, J., NAKAMURA, Y., TOKSÖZ, N., LAMMLEIN, D., DUENNEBIER, F. and DAINTY, A., 1973. Passive seismic experiment. *Apollo 17 Preliminary Science Report*, Chapter 11, NASA SP–330.

LEVINSON, A. A. and TAYLOR, S. R., 1971. *Moon Rocks and Minerals*. Oxford: Pergamon Press.

McCALL, J., 1965. The caldera analogy in selenology. *N.Y. Acad. of Sci.*, **123**, 843–875.

McGETCHIN, T. R. and HEAD, J. W., 1973. Lunar cinder cones, *Science*, **180**, 68–71.

McGILL, G. E., 1971. Attitude of fractures bounding straight and arcuate lunar rilles. *Icarus*, **14**, 53–58.

MASON, R., GUEST, J. E. and COOKE, G., 1976. An Imbrium pattern of graben on the Moon. *Proc. Geol. Assn.*, London, **2**, 161–168.

MILTON, D., BARLOW, B. C., BRETT, R., BROWN, A. R., GLIKSON, A. Y., MANWARING, E. A., MOSS, F. J., SEDMIK, E. C. E., SON, J. VAN and YOUNG, G. A., 1972. Gosses Bluff Impact Structure, Australia. *Science*, **175**, 1199–1207.

MOORE, H., HODGES, C. A. and SCOTT, D. H., 1974. Multiringed basins illustrated by Orientale and associated features. *Proc. 5th Lunar Conf., Geochim. Cosmochim. Acta*, Suppl. 5, pp. 71–100.

MORRIS, E. G. and WILHELMS, D. E., 1967. Geologic map of the Julius Caesar quadrangle of the Moon. *Map I-510, U.S. Geol. Surv. Washington, D.C.*

MUEHLBERGER, W. R., BATRON, R. M., BOUDETTE, E. L., DUKE, C. M., EGGLETON, R. E., ELSTON, D. P., ENGLAND, A. W., FREEMAN, V. L., HAIT, M. H., HALL, T. A., HEAD, J. W., HODGES, C. A., HOLT, H. E., JACKSON, D. E., JORDAN, J. A., LARSON, K. B., MILTON, D. J., REED, V. S., RENNILSON, J. J., SCHABER, C. G., SCHAFER, J. P., SILVER, L. T., STUART-ALEXANDER, D., SUTTON, R. L., SWANN, G. A., TYNER, R. L., ULRICH, G. E., WILSHIRE, H. G., WOLF, E. W. and YOUNG, J. W., 1972. Preliminary geologic investigation of the Apollo 16 site. *Apollo 16 Prelim. Sci. Report*. NASA SP–315, chapter 6, pp. 6–1 to 6–81.

MULLER, P. M. and SJOGREN, W. L., 1968. Mascons: lunar mass concentrations. *Science*, **161**, 680–684.

MURASE, T. and McBIRNEY, A. R., 1970 a. Viscosity of lunar lavas. *Science*, **167**, 1491–1493.

MURASE, T. and McBIRNEY, A. R., 1970 b. Thermal conductivity of lunar and terrestrial igneous rocks in their melting range. *Science*, **170**, 165–167.

MURRAY, B. C., BELTON, M. J. S., DANIELSON, G. E., DAVIS, M. E., GAULT, D. E., HAPKE, B., O'LEARY, B., STROM, R. G., SUOMI, V., and TRASK, N. J., 1974 a. Mariner 10 pictures of Mercury: First results. *Science*, **184**, 459–461.

MURRAY, B. C., BELTON, M. J. S., DANIELSON, G. E., DAVIES, M. E., GAULT, D. E., HAPKE, B., O'LEARY, B., STROM, R. G., SUOMI, V. and TRASK, N. J., 1974 b. Mercury's surface: Preliminary description and interpretation from Mariner 10 pictures. *Science*, **185**, 169–179.

MURRAY, J. B. and GUEST, J. E., 1970. Circularities of craters and related structures on Earth and Moon. *Mod. Geology*, **1**, 149–159.

MUTCH, T. A., ARVIDSON, R., JONES, K., HEAD, J. W. and SAUNDERS, R. A., 1977. *Geology of Mars*. Princeton University Press.

OBERBECK, V. R., 1970. Lunar dimple craters. *Mod. Geol.*, **1**, 161–171.

OBERBECK, V. R., 1975. The role of ballistic erosion and sedimentation in lunar stratigraphy. *Rev. Geophys. Space Physics*, **13**, 337–362.

OBERBECK, V. R. and MORRISON, R. H., 1973. On the formation of the lunar herringbone pattern. *Geochim. Cosmochim. Acta*, **37**, Suppl. 4, 107–123.

OBERBECK, V. R., HÖRZ, F., MORRISON, R. M. and QUAIDE, W. L., 1973. Emplacement of the Cayley Formation. NASA TM X–62, 302, 38p.

OBERBECK, V. R. and QUAIDE, W. L., 1968. Genetic implications of lunar regolith thickness variations. *Icarus*, **9**, 446–465.

OBERBECK, V. R., QUAIDE, W. L. and GREELEY, R., 1969. On the origin of sinuous rilles. *Mod. Geol.*, **1**, 75–80.

OBERBECK, V. R., QUAIDE, W. L., MAHAN, M. and PAULSON, J., 1973. Monte Carlo calculations of lunar regolith thickness distributions. *Icarus*, **19**, 87–107.

O'KEEFE, J. A., 1969. Origin of the Moon. *J. Geophys. Res.*, **74**, 2758–2767.

PETERSON, D. W. and SWANSON, D. A., 1974. Observed formation of lava tubes. *Studies in Speleology*, **2**, 209–222.

PIKE, R. J., 1974 a. Depth/diameter relations of fresh lunar craters: Revision from spacecraft data. *Geophys. Res. Letters*, **1**, 291–294.

PIKE, R. J., 1974 b. Craters on Earth, Moon and Mars: Multivariate classification and mode of origin. *Earth and Planetary Sci. Letters*, **22**, 245–255.

POHN, H. A. and OFFIELD, T. W., 1970. Lunar crater morphology and relative-age determination of lunar geologic units—Part I. Classification. *U.S. Geol. Survey Prof. Paper*, 700–C, pp. C153–C162.

QUAIDE, W. L. and OBERBECK, V. R., 1968. Thickness determinations of the lunar surface layer from impact craters. *J. Geophys. Res.*, **73**, 5247–5270.

QUAIDE, W. L. and OBERBECK, V. R., 1975. Development of mare regolith: some model considerations. (Manuscript.)

READ, H. H. and WATSON, J., 1975. *Introduction to Geology*. Vol. 2: Earth History, Part I. London: Macmillan.

RINGWOOD, A. E., 1970. Origin of the Moon: the precipitation hypothesis. *Earth Planetary Sci. Lett.*, **8**, 131–140.

RINGWOOD, A. E. and ESSENE, E., 1970. Petrogenesis of Apollo 11 basalts, internal constitution and origin of the Moon. *Geochim. Cosmochim. Acta*, **34**, suppl. 1, 769–799.

RODDY, D., 1963. The Flynn Creek crater, Tennessee. In *Shock Metamorphism of Natural Materials* (Eds. French, B. M. and Short, N.), pp. 291–322. Baltimore: Mono Book Corp.

RODDY, D. J., BOYCE, J. M., COTTON, G. W. and DIAL, A. C., 1975. Meteor Crater, Arizona, rim drilling with thickness, structural uplift, diameter, depth, volume and mass-balance calculations. *Proc. 6th Lunar Science Conf.*, *Geochim. Cosmochim. Acta* (suppl. 6), 2621–2644.

SCHABER, G. G., 1973. Eratosthenian volcanism in Mare Imbrium: Source of youngest lava flows. *Apollo 17 Preliminary Science Report*, NASA SP–330, pp. 30–17 to 30–21.

SCHULTZ, P. H., 1976. *Moon Morphology*. Austin: University of Texas Press.

SCHULTZ, P. H. and GAULT, D. E., 1975. Seismic effects from major basin formation on the Moon and Mercury. *The Moon*, **12**, 159–177.

SCHULTZ, P. and GREELEY, R., 1976. Ring moat structures: preserved flow morphology on lunar maria. *Lunar Science VII* (Lunar Science Institute, Houston), pp. 788–790.

SHOEMAKER, E. M., 1960. Penetration mechanics of high velocity meteorites, illustrated by Meteor Crater, Arizona. *International Geol. Congress*, XXI session, Norden, pt. XVIII.

SHOEMAKER, E. M., 1962. Interpretation of lunar craters. In *Physics and Astronomy of the Moon* (Ed. Kopal, Z.), Chapter 8, pp. 283–303. New York: Academic Press.

SHOEMAKER, E. M., 1965. Preliminary analysis of the fine structure of the lunar surface in Mare Cognitum. In *Nature of the Lunar Surface* (Eds. Menzel, W. N. and O'Keefe, J. A.), Chapter 2. Baltimore: John Hopkins Univ. Press.

SHOEMAKER, E. M., BATSON, R. M., HOLT, H. E., MORRIS, E. C., RENNILSON, J. J. and WHITAKER, E. A., 1968. Television observation from Surveyor VII. In *Surveyor VII: A Preliminary Report*. NASA SP–173, pp. 13–81.

SHOEMAKER, E. M. and HACKMAN, R. J., 1962. Stratigraphic basis for a lunar time scale. *The Moon*, Symp. 14 International Astronomical Union (Eds. Kopel, Z. and Mikhailov Z. K.), pp. 289–300. London: Academic Press.

SHORT, N. M., 1975. *Planetary Geology*. New Jersey: Prentice-Hall Inc.

SHREVE, R. L., 1966. The Sherman landslide, Alaska. *Science*, **154**, 1639–1643.

SODERBLOM, L. A., WEST, R. A., HERMAN, B. M., KREIDLER, T. J. and CONDIT, C. D., 1974. Martian planetwide crater distributions: Implications for geologic history and surface processes. *Icarus*, **22**, 239–263.

SPURR, J. E., 1944 (Vol. 1), 1945 (Vol. 2), 1948 (Vol. 3) and 1949 (Vol. 4). *Geology Applied to Selenology*. New Hampshire: The Romford Press.

STROM, R. G., 1964. Analysis of lunar lineaments, I: Tectonic maps of the Moon. *Comm. Lunar Planet. Lab. Univ. Arizona, No. 39*, pp. 205–216.

STROM, R. G., 1965. Interpretation of Ranger VII records. *Jet Prop. Lab. Report* **32**, Chapter III.

STROM, R. G., 1971. Lunar mare ridges, rings and volcanic ring complexes. *Mod. Geol.*, **2**, 133–157.

STROM, R. G. and FIELDER, G., 1970. Multiphase eruptions associated with Tychus and Aristarchus. *Lunar Planet. Lab. Comm. No. 150*, pp. 235–288.

STROM, R. G., TRASK, N. J. and GUEST, J. E., 1975. Tectonism and volcanism on Mercury. *J. Geophys. Res.*, **80**, 2478–2507.

STUART-ALEXANDER, D. and HOWARD, K. A., 1970. Lunar maria and circular basins—A review. *Icarus*, **12**, 440–456.

STUART-ALEXANDER, D. E. and WILHELMS, D. E., 1974. The Nectarian system, a new lunar time stratigraphic unit. *J. Res. U.S. Geol. Surv.*

TAYLOR, S. R., 1975. *Lunar Science: A Post Apollo View*. New York: Pergamon.

THOMPSON, T. W. and DYCE, R. B., 1966. Mapping of lunar radar reflectivity at 70 centimeters. *J. Geophys. Res.*, **71**, 4843–4853.

THOMPSON, T. W., MASURSKY, H., SHORTHILL, R. W., TYLER, G. L. and ZISK, S. H., 1974. A comparison of infrared, radar and geologic mapping of lunar craters. *The Moon*, **10**, 87–117.

TRASK, N. J. and GUEST, J. E., 1975. Preliminary geologic terrain map of Mercury. *J. Geophys. Res.*, **80**, pp. 2461–2477.

TRASK, N. J. and McCAULEY, J., 1972. Differentiation and volcanism in the lunar highlands: Photogeologic evidence and Apollo 16 implications. *Earth Planet. Sci. Letters*, **14**, 201–206.

WALKER, G. P. L., 1972. Lengths of lava flows. *Phil. Trans. Roy. Soc. London*, **274**, 107–118.

WARNER, J. L., 1975. Lunar thin section educational package (text). Houston, Texas: NASA Johnson Space Center.

WHITAKER, E. A., 1972. Mare Imbrium flows and their relationship to colour boundaries. *Apollo 15* Preliminary Science Report. NASA SP–289, pp. 25–83 to 25–84.

WILHELMS, D. E., 1972. Geologic mapping for the second planet. *Interagency Report: Astrogeology*, **55**, 36p.

WILHELMS, D. E. and DAVIS, D. E., 1971. Two former faces of the Moon. *Icarus*, **15**, 368–372.

WILHELMS, D. E. and MCCAULEY, J. F., 1971. Geologic map of the near side of the Moon. *Geologic Atlas of the Moon*, U.S. Geol. Survey 1:5 000 000 Map I—703.

WOOD, C. A. and HEAD, J. W., 1975. Pre-mare material of possible volcanic origin. In *Origins of Mare Basalts and their Implications for Lunar Evolution*, pp. 192–193. Houston: Lunar Science Institute.

Index of Names

Subject Index

THE WYKEHAM SCIENCE SERIES

THE WYKEHAM ENGINEERING AND TECHNOLOGY SERIES

All orders and requests for inspection copies should be sent to the appropriate agents. A list of agents and their territories is given on the verso of the title page of this book.

†*(Paper and Cloth Editions available.)*